LUXE

深圳市创扬文化传播有限公司 编
赵 欣 译

VILLA DESIGN COLLECTION II

别墅 奢华典藏 II

大连理工大学出版社
Dalian University of Technology Press

图书在版编目(CIP)数据

别墅：奢华典藏. 2：汉英对照 / 深圳市创扬文化传播有限公司编；赵欣译. — 大连：大连理工大学出版社，2011.5
ISBN 978-7-5611-6133-3

Ⅰ.①别… Ⅱ.①深… ②赵… Ⅲ.①别墅—建筑设计—作品集—世界 Ⅳ.①TU241.1

中国版本图书馆CIP数据核字（2011）第052946号

出版发行：大连理工大学出版社
（地址：大连市软件园路80号　邮编：116023）
印　　刷：利丰雅高印刷（深圳）有限公司
幅面尺寸：280mm × 280mm
印　　张：28
插　　页：4
出版时间：2011年5月第1版
印刷时间：2011年5月第1次印刷
策　　划：袁　斌
责任编辑：刘　蓉
责任校对：王　琴
封面设计：温广强

ISBN 978-7-5611-6133-3
定　　价：328.00元

电 话：0411-84708842
传 真：0411-84701466
邮 购：0411-84703636
E-mail: designbooks_dutp@yahoo.cn
URL: http://www.dutp.cn

如有质量问题请联系出版中心：（0411）84709246　84709043

WARHOL LIVE
HERB RITTS

Contents 目录

地点	面积	设计师	设计公司
深圳	525平方米	刘伟婷	刘伟婷设计师有限公司

Time Garden Villa of Six Floors (Building A3)

时代花园6层别墅（A3栋）

This project integrates the British and Indian royal feelings. In the living room, the strong British style and classical palace style are exposed completely, and the skillful art on the furniture reveals the Victorian architecture style. The living room is decorated with a giant crystal background wall of Baroque style, and the delicate carvings attract people's eyes.

Walking slowly down along the carved stairs to the basement is the banquet hall. The gorgeous and elegant scene leaps to the eyes, and the carpet of Indian style is an important decoration in the house, with its delicate workmanship. Next to the hall is the British-style billiards room. The guests who like the quiet activities are invited to the second floor underground, including movie and yoga. The three-dimensional audio and video equipment is coupled with a big screen, just like holding a small premiere, so movie fans get together here. The simply designed yoga room reveals a natural and peaceful feeling, and the pool filled with carved patterns leads people into the silence of the soul.

The master bedroom creates a romantic style full of noble flavor. The complex decoration, carvings, stable headboard with beautiful lines, carvings on arched parquet board; all those careful design elements can raise the owner's status and taste. The artistic carpet is the essence of the master bedroom, making the British and Indian royal feelings clearly printed here.

未使用
已使用

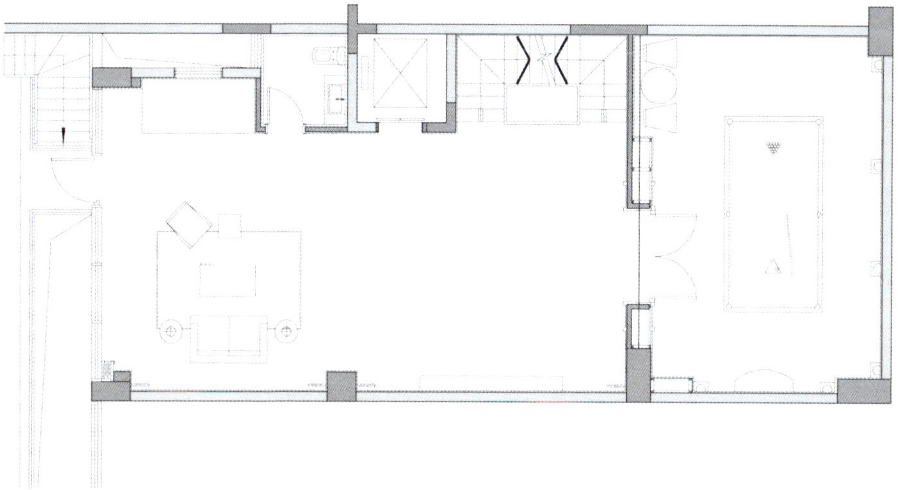

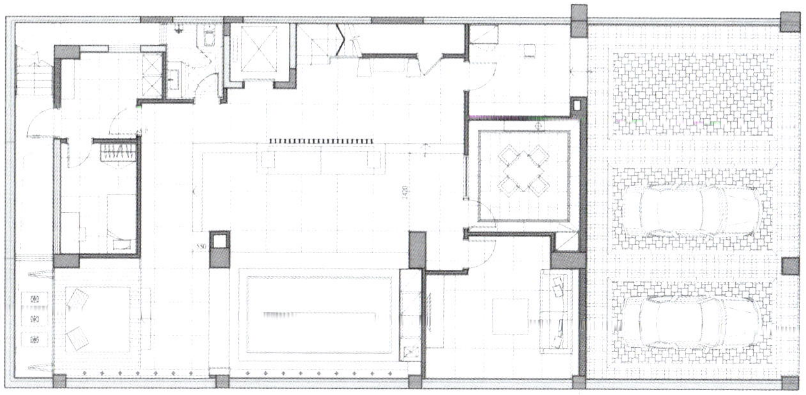

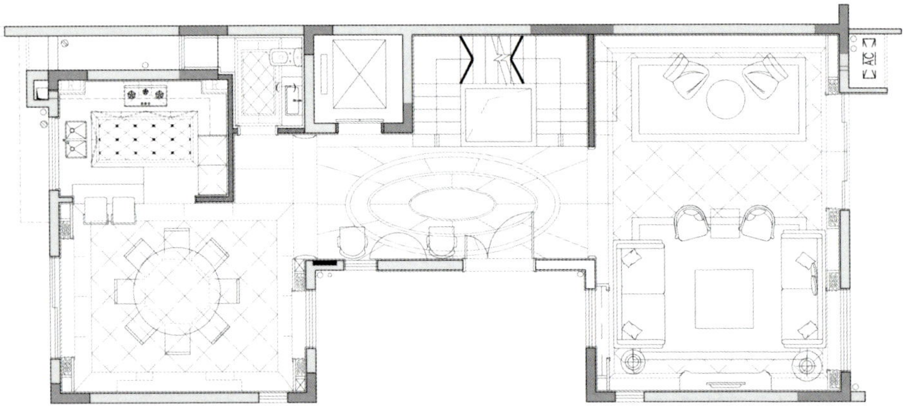
AC

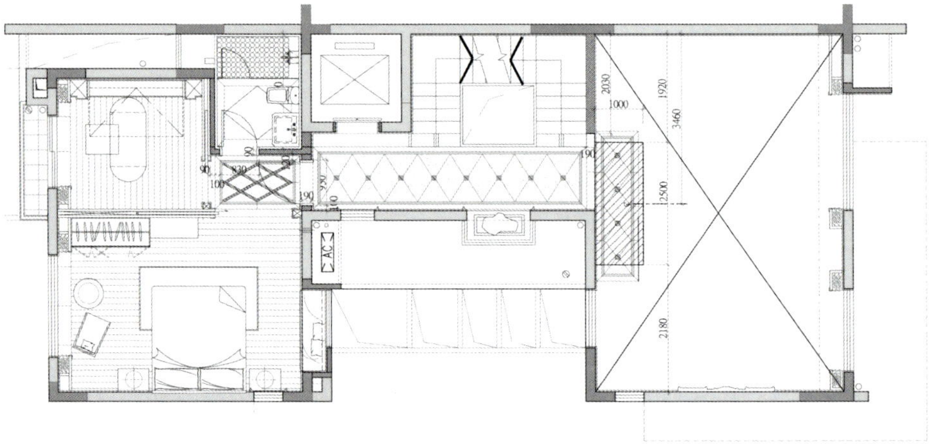
AC
2030
1920
1000
3460
2500
2180

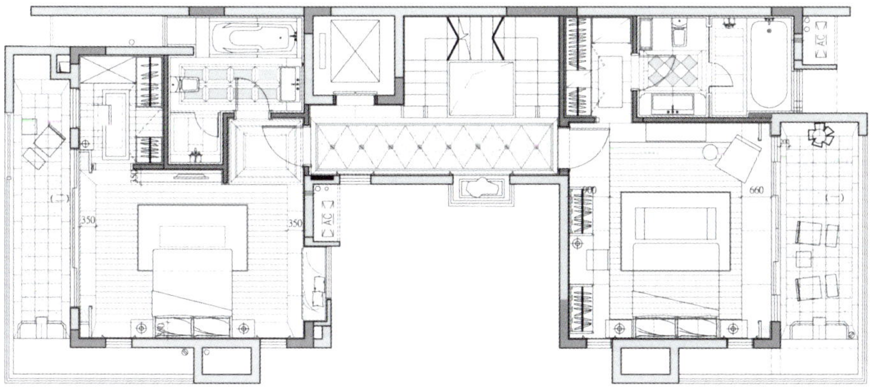

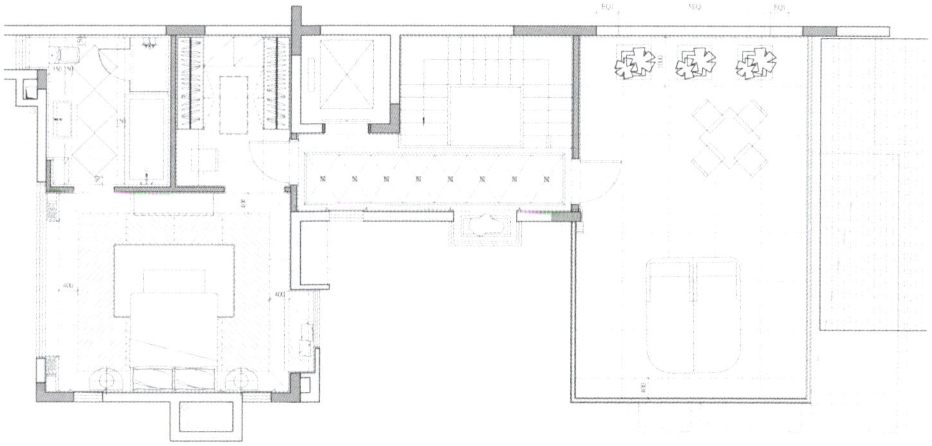

本案融合了英国和印度的皇室风情。在客厅中，浓浓的英伦风情及古典宫廷气派表露无遗，家具的巧手技艺流露出英国维多利亚式建筑风格。客厅中挂着一幅英国巴洛克式巨型水晶幕墙，细腻的雕花将客人的目光汇聚于此。

沿着雕花楼梯缓缓走向地下一层便是宴会厅，华丽典雅的情景摄入眼帘。充满印度风情的地毯是家里重要的装饰品，其细腻的工艺令人赞叹不已。宴会厅旁边为英式桌球室。而喜爱安静的客人被邀请到地下二层，电影欣赏、瑜伽运动都包含其中，立体影音配上大银幕，像举行小型首映礼般，令电影爱好者汇聚一堂。简洁设计的瑜伽室流露出自然、平和的感情，充满雕刻花纹的水池，带人走进静默的心灵。

主卧室营造出浪漫但极具贵族风范的气派，繁复的装饰、雕花、线条优美且稳重的床头板、圆拱形的镶花板雕刻，精心的设计往往能抬高主人的身份和品位。一张艺术气质浓厚的地毯亦是主卧室的精髓，让皇室风范清晰地印在其中。

面积	设计师	设计公司
293平方米	刘伟婷	刘伟婷设计师有限公司

Oriental Charm of Mandarin Hotel, Apartment A1
文华酒店之东方魅力A1户型

For the square and tidy structure, the designer chooses modern-style quadrangle dwelling design, which means the mansion's central axis is the main building structure. In order to highlight the family style, the designer places door-height beams and pitch-black tune at the entrance to show the pragmatic, honest and simple atmosphere, surrounded by galleries, and uses of Chinese red, rose gold, ebony and other valuable and auspicious colors to show rich culture.

In the dining room, the simple lines, magnificent and towering effect of hollowed design, and the wall with concave-convex natural travertine are gorgeous but not flashy. The chairs imitating Ming Dynasty hat styled with high backs and the long-shape crystal dining table with eight seats create a unique fashion, and the exaggerated hanging lighting is coupled with abstract paintings of modern Chinese art style to create equally satisfactory results with different approaches. In the bedroom area on the second floor, the headboard is designed with a curtain with golden-embroidery auspicious patterns of flowers, coupled with soft and smooth silk bedding and elegant velvet bench, showing a romantic oriental expression.

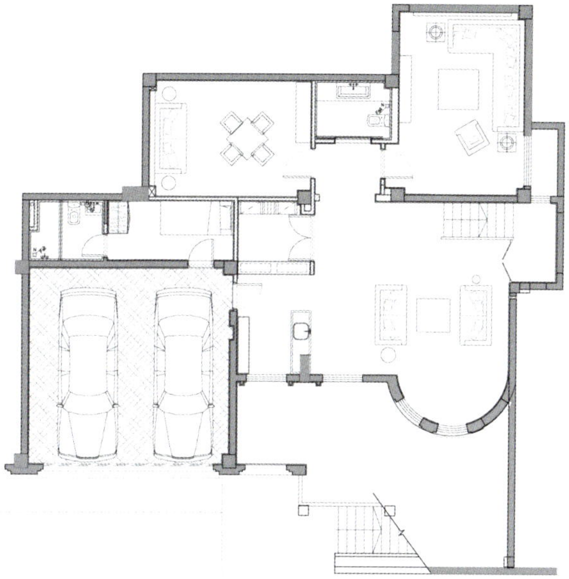

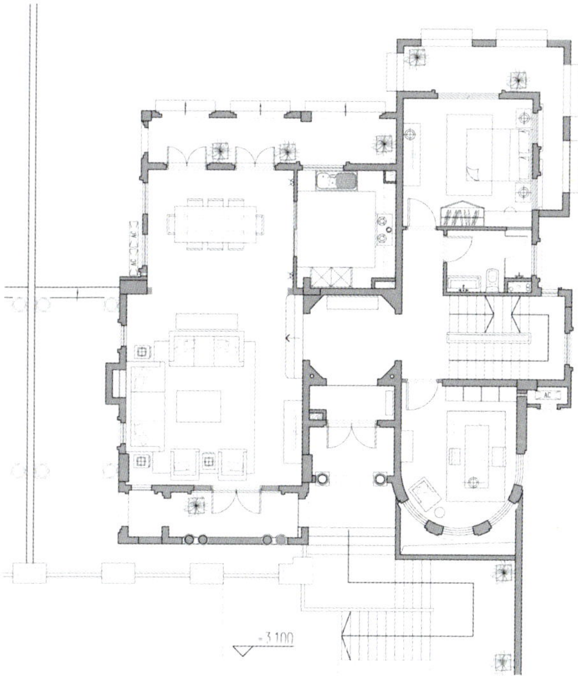

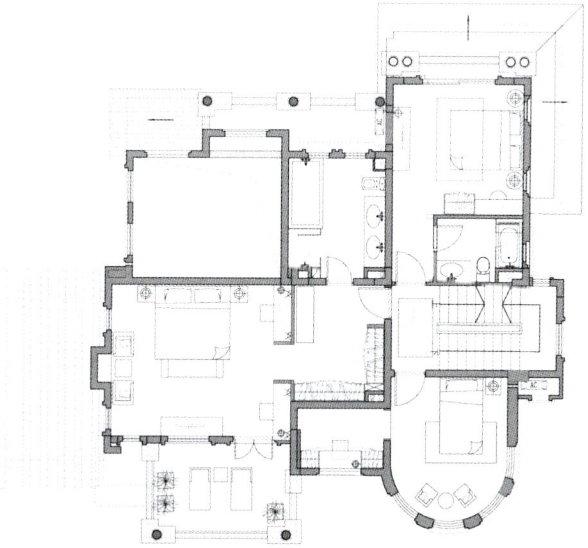

由于结构方正整齐，设计师采用了现代式的四合院设计，即以大宅的中轴线为主体。为突出家族的气派，入口大门设计了骑门梁，乌黑的色调表现出沉实敦厚，周围以廊相接，同时采用中国红、玫瑰金及乌檀木色等吉祥的颜色，流露丰厚的文化底蕴。

餐厅线条简约，中空设计产生壮丽高耸的效果，天然窿石的凹凸直坑饰面墙身，华丽而不艳俗。仿明朝高身独背板官帽餐椅与长形水晶玻璃八人餐桌营造出别具一格的时尚气息。夸张的吊挂灯饰配衬抽象的后现代中国艺术风格挂画，二者有着异曲同工之妙。登上二层的卧室区，绣金的吉祥花卉图案屏障床头板配上柔软光滑的丝绢寝具和典雅的丝绒床尾椅，流露出浪漫的东方神采。

面积	设计师	设计公司	主要材料
400平方米	史林艳	香港神采设计建筑装饰总公司 宁波神采装饰设计工程有限公司	橡木饰面、米白洞石、深蓝大理石、仿古砖、仿古马赛克

Nobletimes No.48 Mexican Style

香蜜湖48#墨西哥风情

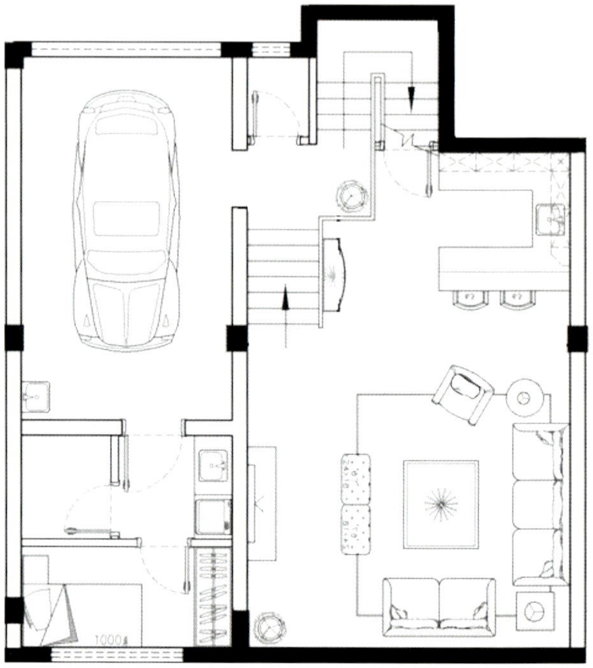

The design fully absorbs the characteristics of Mexico style, implementing the simple and elegant way, and the designer makes a big fuss about the collocation of colors, where the whole style is simple and unrestrained, free and easy, full of strong exotic flavor.

Because the house is designed for young people, the colors collocation is more colorful and rich. The designer boldly uses the main color of blue, white and yellow, and applies them to every room and every facade of each floor. Take the living room as an example, at the entrance you can see a large Mediterranean hand painting in blue, where sea and sky merge into one, bringing you unlimited imagination. The introduction of yellow is reflected on the opposite sofa, while the furniture and dome are designed with natural wood color, coupled with iron droplights, creating an elegant and luxurious atmosphere. The use of colors is also extended into the master bedroom, study, children's room, basement and other rooms, the colors of blue, white and yellow can be seen everywhere, supplemented by colorful decorative accessories, full of vitality and showing the consistent characteristic of the whole design. Furniture and ceiling are mainly simple shapes with natural wood color, pursuing the natural flavor, and the design features a certain sense of history deposition, showing a simple and luxurious concept.

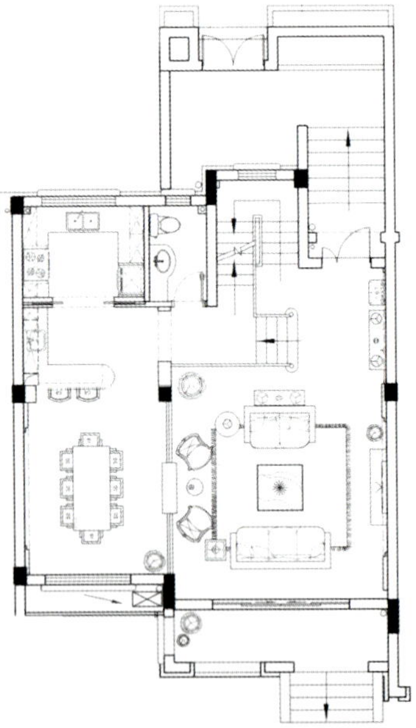

本案在设计上充分吸纳墨西哥风情的特质，贯彻简约大方的路线，在颜色的搭配上做足文章，整体风格既简单奔放、自由洒脱，又充满了浓郁的异国风情。

因为居住者是年轻人，所以在色彩搭配上更加丰富艳丽。设计师大胆采用蓝、白、黄为主色系，并将其应用到每一个房间、每一层立面中。以客厅为例，步入玄关即可看到一幅满目天蓝的地中海手绘画，水天相接，浑然一体，带给人无限遐想。黄色的运用则体现在对面的沙发上，家具、穹顶则使用天然原木色，搭配铁艺吊顶灯，营造优雅奢华的空间氛围。这一色彩运用也被延伸到主卧室、书房、儿童房、地下室等各个空间，随处可见以蓝、白、黄色系为主，色彩绚丽的饰品装饰为辅的色彩搭配手法，充满了活力，体现了作品整体一致的内在特征。家具和吊顶以天然木质原色为主，造型简洁，追求本色，有一定的历史积淀感，体现了简约奢华的理念。

DYNASTY

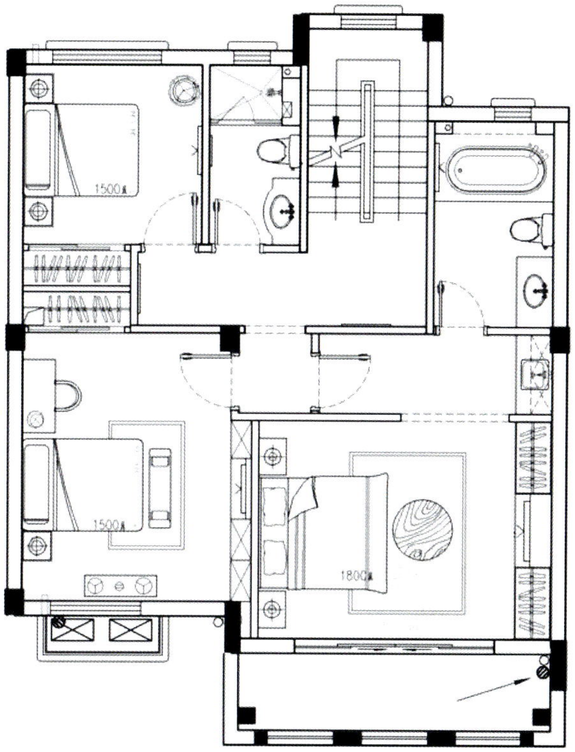

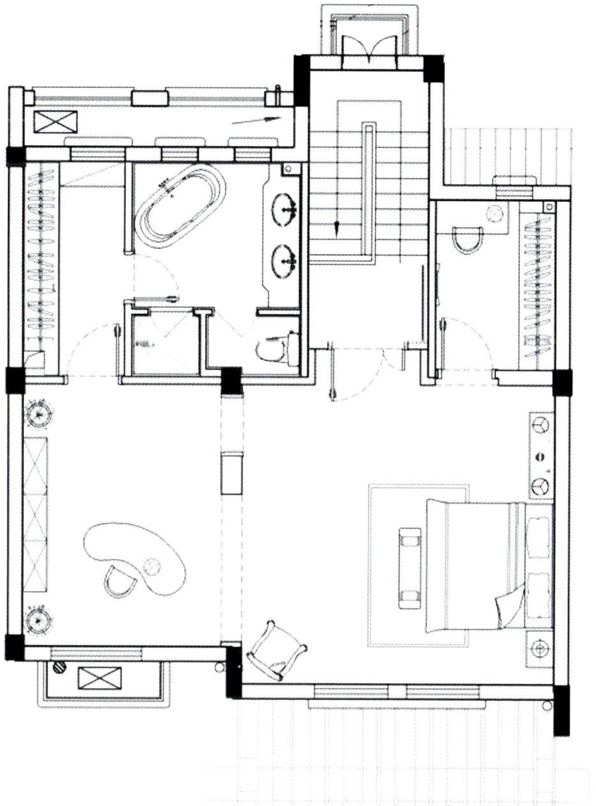

地点	面积	设计师	设计公司	主要材料	摄影师
惠州	700平方米	薛峰、朱晓鸥、王君明、娄欣	寅界（国际）建筑室内设计	各种云石、柚木饰面、绿可木、手扫漆、钢琴漆、墙纸	郭衍军

Poly Landscape City Villa, Huizhou

惠州保利山水城别墅

The designer applies the European luxurious style, and endows a resplendent romance to the whole space. The double layered spatial space of the living room provides plenty room for the hanging curtains, luxurious drop light, and large sized hanging ornaments. The adoption of the ceiling mirrors in the dining room not only breaks the suppression and oppressiveness of the ceiling, but further serves to promote the spatial feeling, making the whole space stretch freely in the whole space and accomplishing the transformation of the space smoothly. The luxurious furniture is rich in details, with the complicated moldings match perfectly with the sober and splendid colors, such as purple, gold and gray, and brings out the best in each other. The viewing balcony that extends into the interior space provides an ideal place for gathering with family and friends.

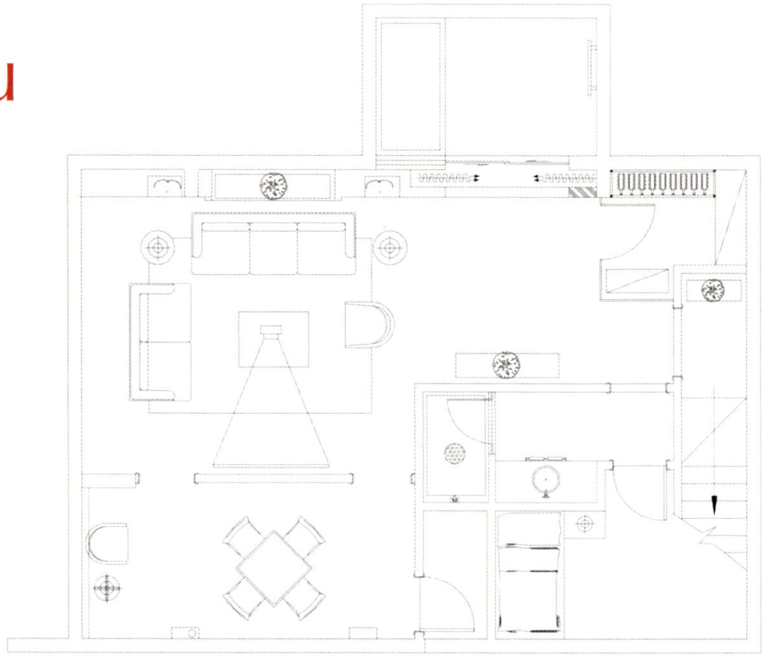

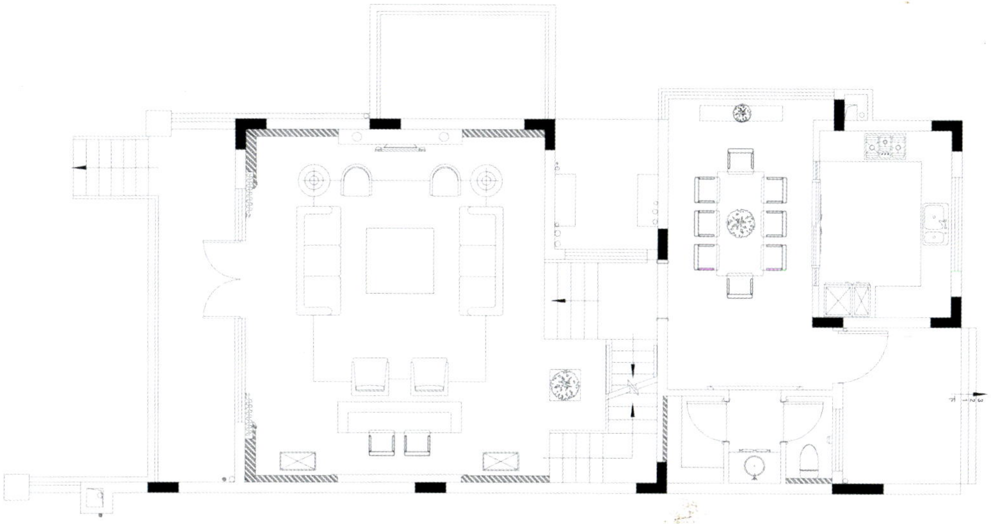

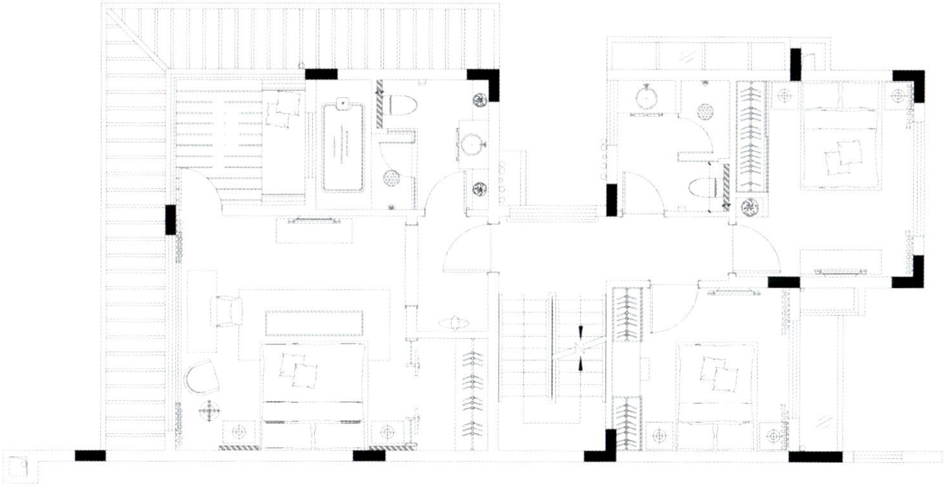

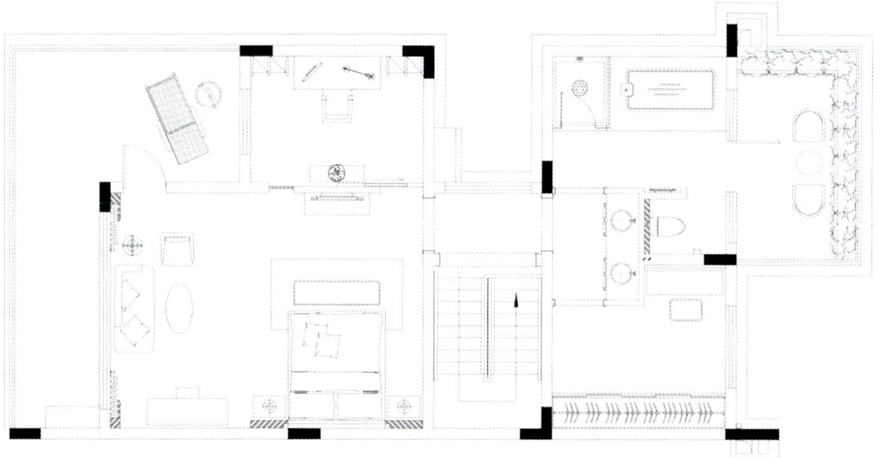

本案的设计师运用了欧式奢华的设计风格，使整个空间洋溢着华丽浪漫的情调。客厅双层的挑空空间给高垂的窗帘、奢华的吊灯和大幅的挂饰以广阔的施展空间。餐厅大花镜面的运用，不仅破除了天花的压抑和沉闷，更起到了推进空间“剧情”的作用，使整个空间自在地延伸，流畅地完成了空间的转换。奢华的家具极富细节，繁复的线脚与紫色、金色、灰色等稳重而华丽的色彩搭配相得益彰。延展室内的生活景观阳台，为家人或朋友聚会提供了理想的场所。

地点	面积	设计师	设计公司	主要材料
重庆	280平方米	冉旭、屈慧颖	重庆旋木室内设计有限公司	水晶黑花岗石、黑色显纹漆、墙纸、木丝水泥板

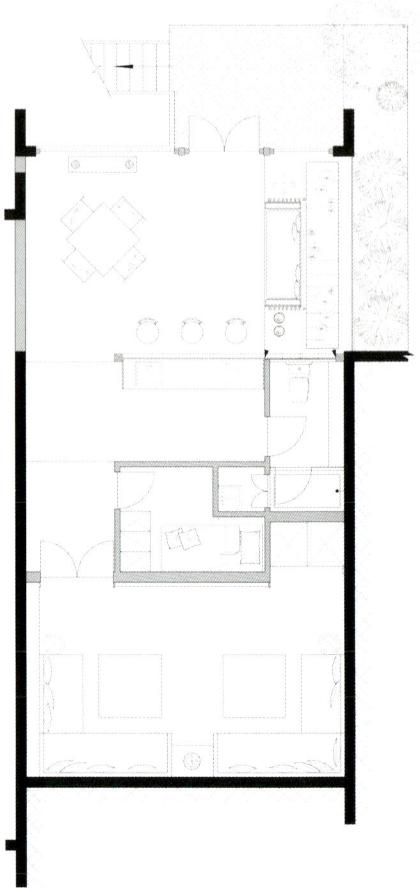

Chongqing Xiaoquan Forest & Sea Villa

重庆小泉森林海别墅

On the keynote of oriental Zen living style, the designers combine the architectural style, geographical surroundings and other objective conditions and grasp the core elements of modern housing to poetically integrate the traditional elements into the modern living space. Low-key colors are used in the design to eliminate the complicated details and create the best integration of the villa and nature.

The living room's TV and kitchen are hidden behind the cream white linen with ink patterns when not in use. Mirror-like black ground, cement countertops and walls are simple and appealing. The shadows of the trees blow in the breeze, casting graceful shadows on the brick-red thick TV wall, mild and elegant. The bed in the master bedroom is right across the valley in its face, and the black leather round window pulls the beauty of the mountain into the interior space, creating a quiet space blending the nature and the living.

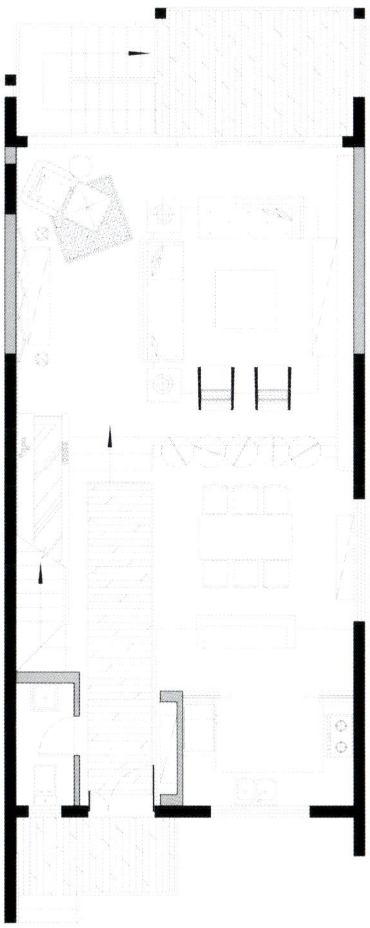

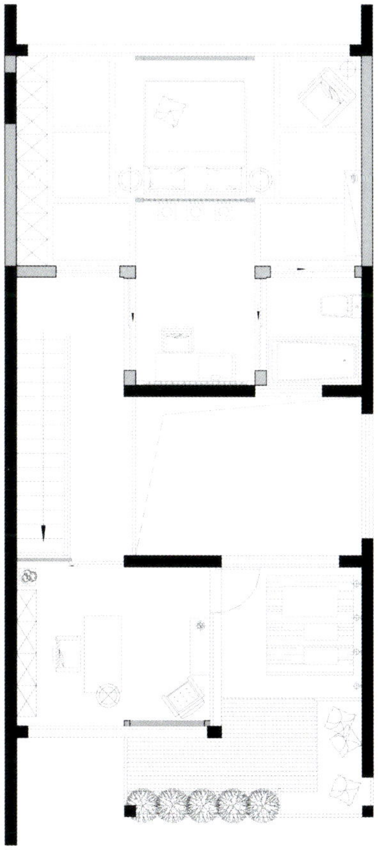

设计师结合建筑风格、地理环境等客观条件，把握现代住宅核心要素，在东方禅意生活的基调上，让传统的元素诗意地融入现代生活空间。设计上使用了低调的色彩，摒弃了繁复的细节，使居室与自然产生最佳融合。

米白色亚麻布上的水墨图案，将客厅的电视和厨房在不使用时隐于其后。如镜的黑色地面、水泥的台面和墙面质朴动人。厚实的砖红色电视墙，微风吹来时，树影在墙面上投射出曼妙的光影，婉转清丽。主卧室的大床正对迎面的山谷，黑色皮质的圆形窗洞把山中美景巧妙引入室内，营造出自然与生活交融的宁静空间。

地点	面积	设计师	设计公司	主要材料
中国台湾	426平方米	吴玟灿	夏川设计/夏川榭环境艺术设计有限公司	石材、铁件、夹纱玻璃、实木、壁纸、裱皮格、人造石

Zhou's Residence, Museum Park Road, Taizhong

台中市美术馆园道周公馆

The designer shows the best and most beautiful side of the space with a generous layout, skillfully using the vertical moving lines of each floor and carefully planning out an amazing variety of functions on life, entertainment, and leisure. The design endows the residence with great charm, especially the respond of inside and outside environments, deep and quiet scene, sense of levels echoing each other in front and behind, and the breathtaking momentum and long-lasting beauty shown out through the clever use of material features and proportion. At the same time, it gives the space temperament and rich culture, just like a grand lyrical epic rooted in minds.

The sense of openness extends to all directions with the coherent dynamic lines, making people deeply feel the broad bearing of mansion upon entering, in particular, the designer skillfully uses the original height of the space, and unifies and justifies a variety of facade shapes serving as different medium in the public spaces to the skyline, including the main and vice walls in the living room, glass screen with clamped yarn, thick sliding doors and other elements. The horizontal and vertical lines full of strength are introduced to show the Zen's quiet attitude to a nicety, delicately outlining modern space with distinctive levels and landscapes in each part.

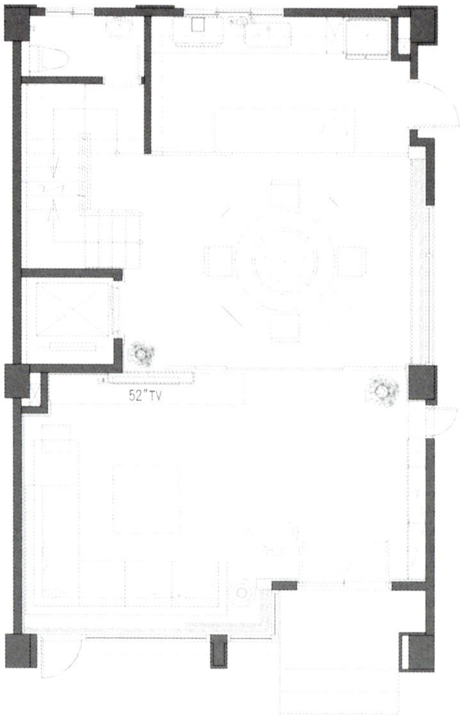

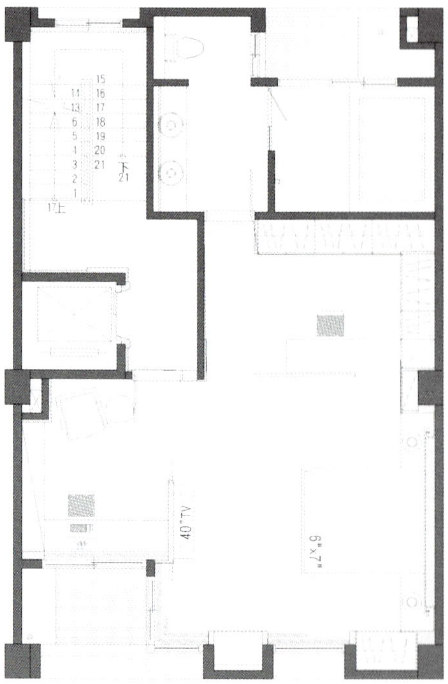

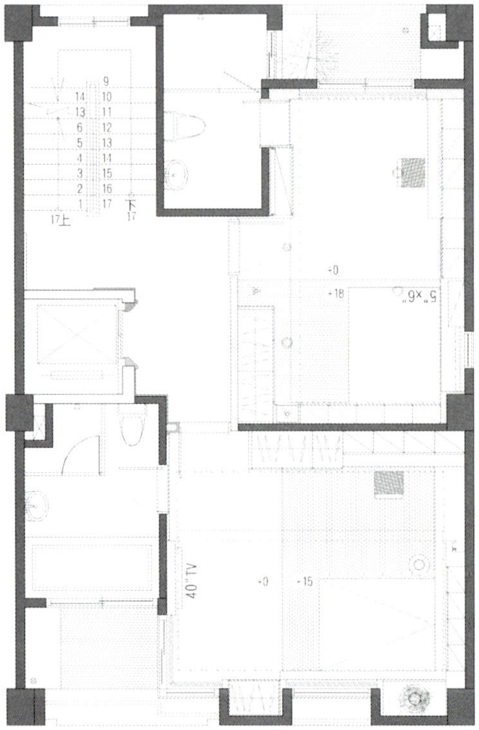

40"TV

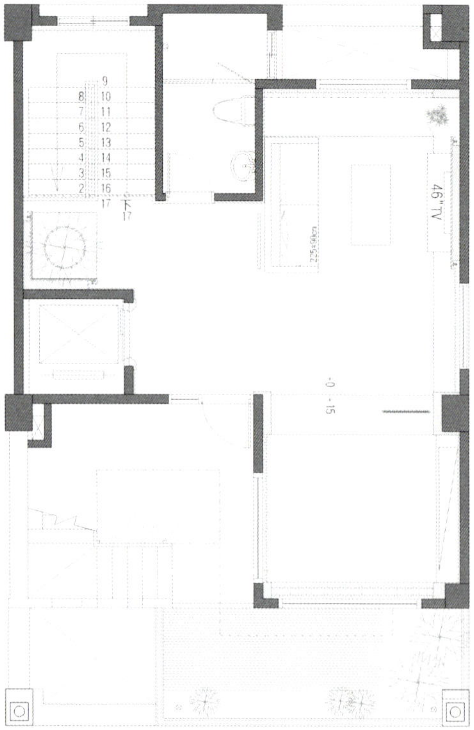

46"TV

设计师以大气的布局手法，融合出空间最美的一面。设计善用楼层垂直动线连接的特征，将空间细心规划出令人叹为观止的生活、娱乐、休闲机能，尤其是环境内外的呼应、幽邃的景深、前后呼应的层次感以及善用材质特性与体量比例所呈现出的磅礴气势与恒久美感，为这座大宅增添了仿若天成的情韵，同时赋予其气质和丰富的人文底蕴，如同一首深植人心的壮阔抒情史诗。

一进门的开阔感随着一气呵成的机能动线向八方延展，让人第一时间就能深刻感受到大宅应有的恢弘气度。特别是设计师利用了先天充裕的室内高度，将担任公共空间不同介质的各种立面造型，包括客厅主副墙、夹纱玻璃隔屏、厚实对开拉门等元素，统一拉齐至等高天际线处理，通过富有力道的水平垂直线条，在洋溢着恰到好处的禅味的宁静致远中，细腻地勾勒出层次鲜明且处处皆有风景的现代空间。

地点	面积	设计师	设计公司	主要材料
宁波	300平方米	朱森钥	宁波市鄞州风尚装饰设计工程有限公司	菠萝格实木、阿曼米黄大理石、意大利蜜蜂砖、美家墙纸、布艺、软包、床品、悦色家具、工艺品

Zi Ting Garden

紫汀花园

This project adopts mix-and-match style which combines both Chinese and Western styles, and takes on a harmonious beauty in clashes. The neoclassicism is benefited from the presentation of mix-and-matching, which not rigidly adheres to the traditional forms, but mixes the extracted elements into the modern styled residence, making the character of the residence much richer.

In this house, furniture with elements of diverse styles exists together, such as furniture of Ming and Qing dynasties, which are made of fine qualities and exquisite craftsmanship, the Chinese lacquer wood furniture in pure or rich colors, cloth sofas of graceful lines, and the European styled iron lamps that inherit Western cultures. All the elements in the household are full of vigorous vitality because of the rich and delicate qualities, including the colors, patterns, materials, brightness and softness. In such a perfect design, either the western or oriental, modern or classical styles are all naturally and perfectly combined.

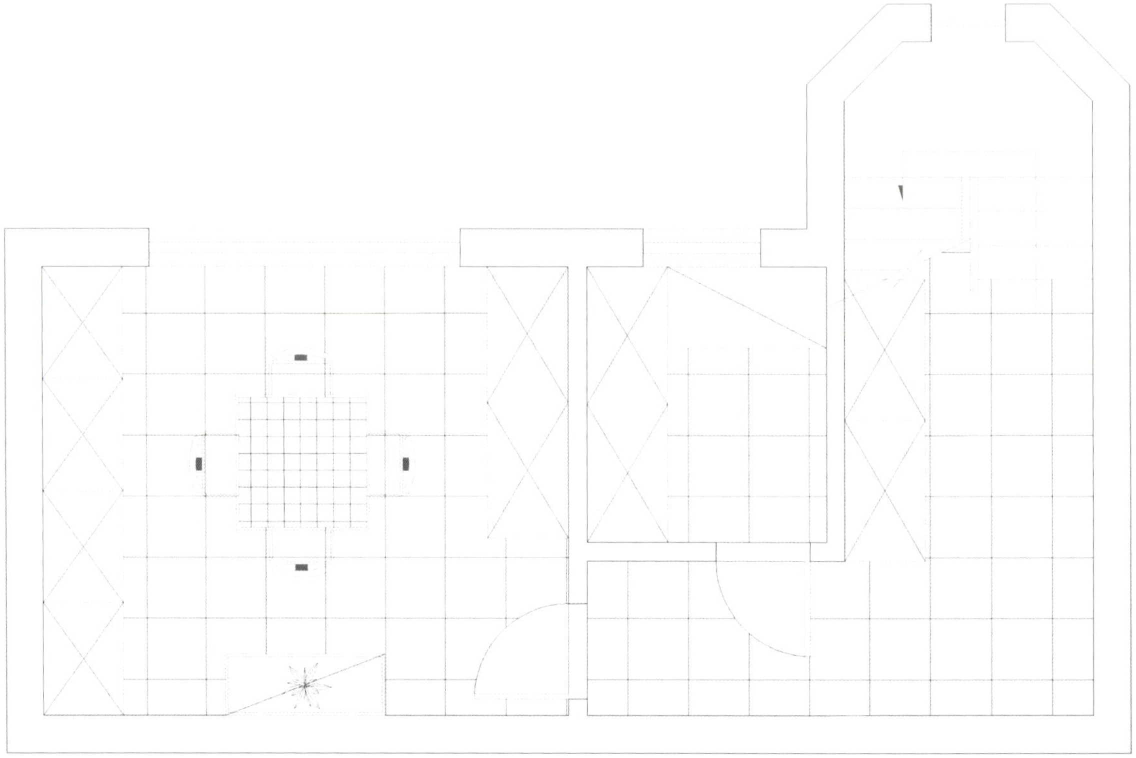

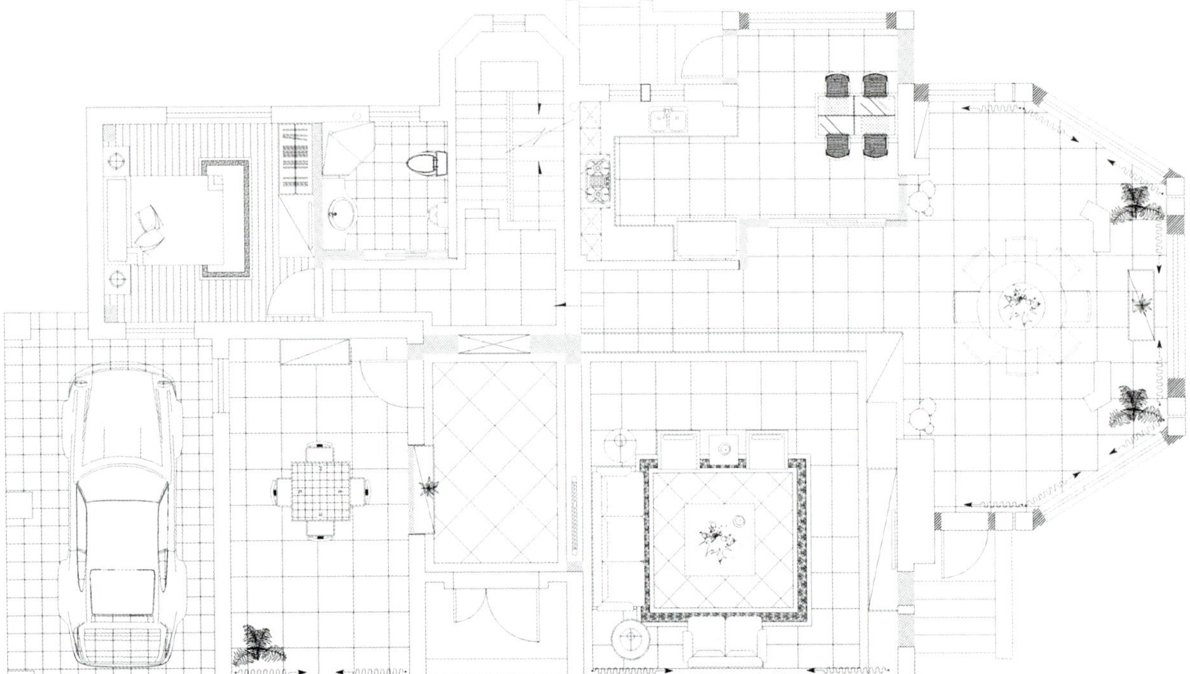

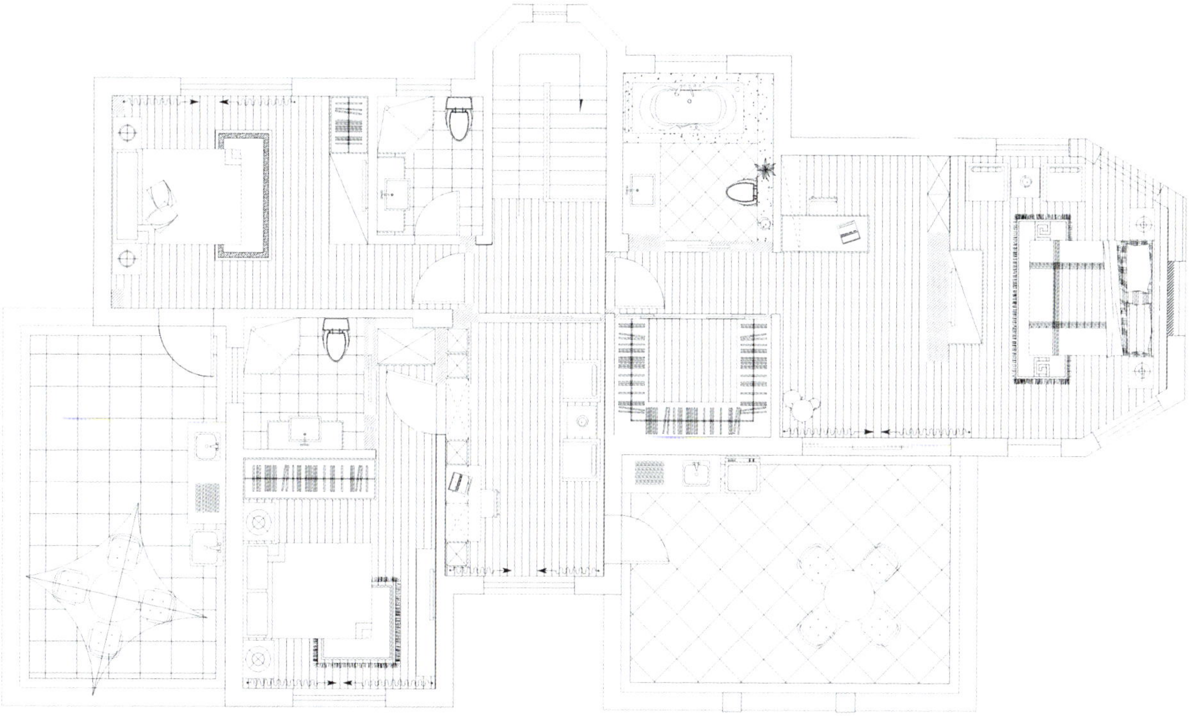

本案采用了中西结合风格的混搭手法，体现碰撞中的和谐之美。新东方风格得益于混搭手法的展示，设计不再拘泥于传统形式，而是将提取的元素融合到现代家居中，使家居的性格更加丰富。

家居内共存着材美工精、典雅简朴的明清家具，纯正单色或五彩缤纷的中式漆木家具、线条柔美的布艺沙发、传承西方文化底蕴的欧式铁艺灯具等不同风格的元素。色彩、图案、材质、亮度、柔度……家居中的一切元素都因丰富细腻的质感而充满蓬勃的生命力。在恰到好处的设计中，西方抑或东方、现代抑或古典都自然地、恰到好处地融合在一起。

山水有清音

Charmmy Kitty

地点	面积	设计师	设计公司	主要材料
哈尔滨	451平方米	万文拓、彭秋波	深圳市品源装饰工程有限公司	软包、地毯、镜面、大理石

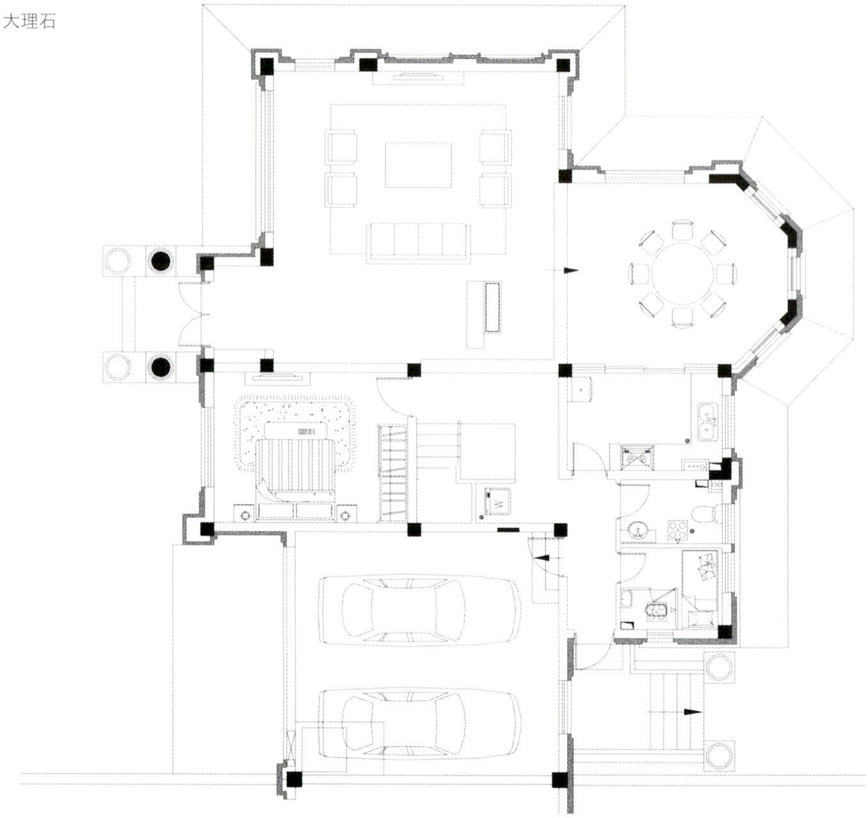

Modern Style of "Black and White Image"

"黑白映像" 现代风

In the project, the designers strive to interpret an easy and healthy modern new life with simple and natural brush work. The whole style is warm and harmonious. The designers choose tough materials, bright colors, soft and stylish furniture, and elegant accessories to endow the space with plain, neat and bright expression.

The living space features adequate lighting, but the original separation is slightly unreasonable, so the designers retain the general spatial structure of the original architecture and make a few changes on the separation to make the use of the spatial structure more rational. The designing elements are capable, and the arrangement of primary and secondary levels is reasonable, forming a good relation of series and showing a stylish and elegant attitude of life.

URBAN DESIGN
FOR AN URBAN CENTURY

本案力求以简洁、自然的笔触演绎轻松健康的时尚新生活，整体风格温馨和谐。设计师以硬朗的材质和明快的色调以及柔美时尚的家具和雅致的配饰，赋予空间素净明亮的神采。

本居住空间采光充足，原始分隔稍有不合理，设计中保留了原建筑的大致空间结构，在分隔上做了些许改动，使空间结构运用更加合理。设计元素干练，主次层次节奏合理，形成了良好的串联关系，表达出时尚高雅的生活态度。

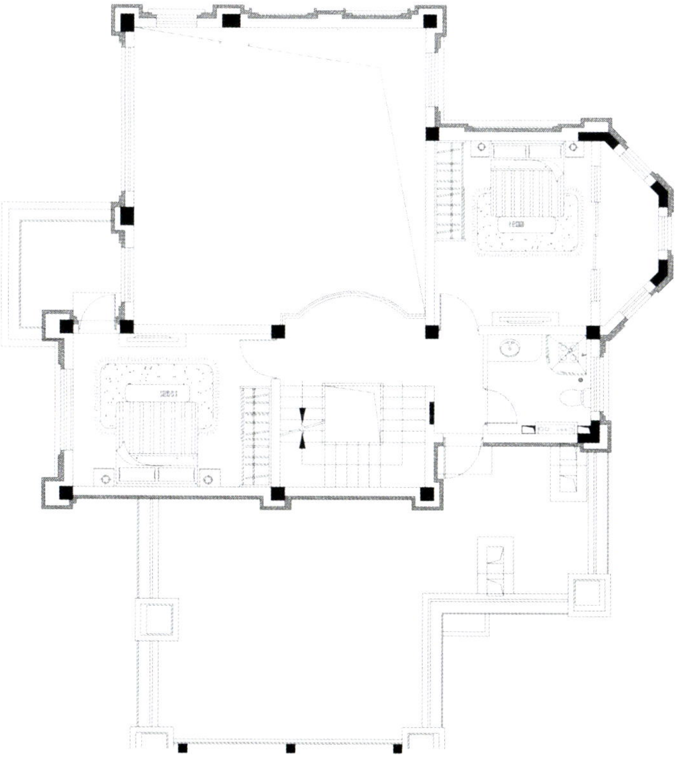

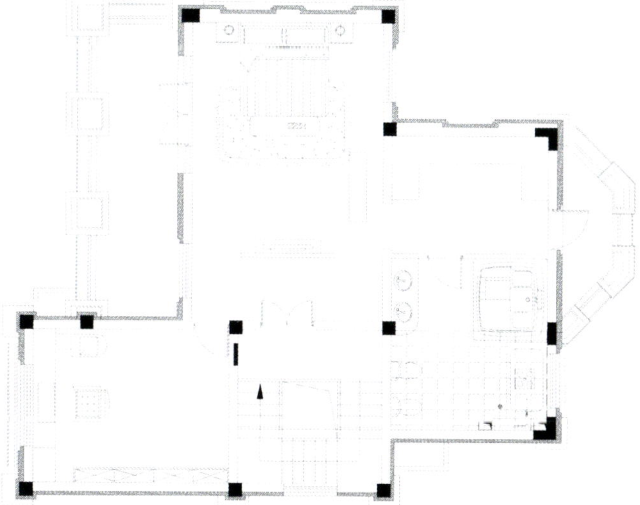

面积	设计师	设计公司	主要材料	摄影师
298平方米	范义峰、黄婷婷	上海天鼓装饰设计有限公司	石材、墙纸、皮质软包、镜面	范义峰

Charming France

迷情法兰西

The project is French style with bright and open space. The designers adopt sophisticated elegant and soft romance as the main tone of the design, highlighting a sweet and romantic atmosphere of life.

On the first floor, both sides of the entrance are added with pillar-type openings, enhancing the connection between facade details and each functional space. On the second floor, the lovely daughter room is filled with all kinds of toys, and the white moldings and soft decoration present the low-key luxurious, which realize the hostess's dream of princess. The piano room on the second floor is just next to the daughter room, allowing the daughter show her art genius better. On the third floor, the master bedroom is designed with soft lines, elegant leather soft decoration and silver foil veneer, fully showing the French romantic. The graceful crystal lighting makes the room smarter. The basement is designed mainly for entertainment, and it is also a place for the hostess to show her excellent cutting skills, allowing the hostess's inspiration fully released here, the scene of daily life features a sweet atmosphere.

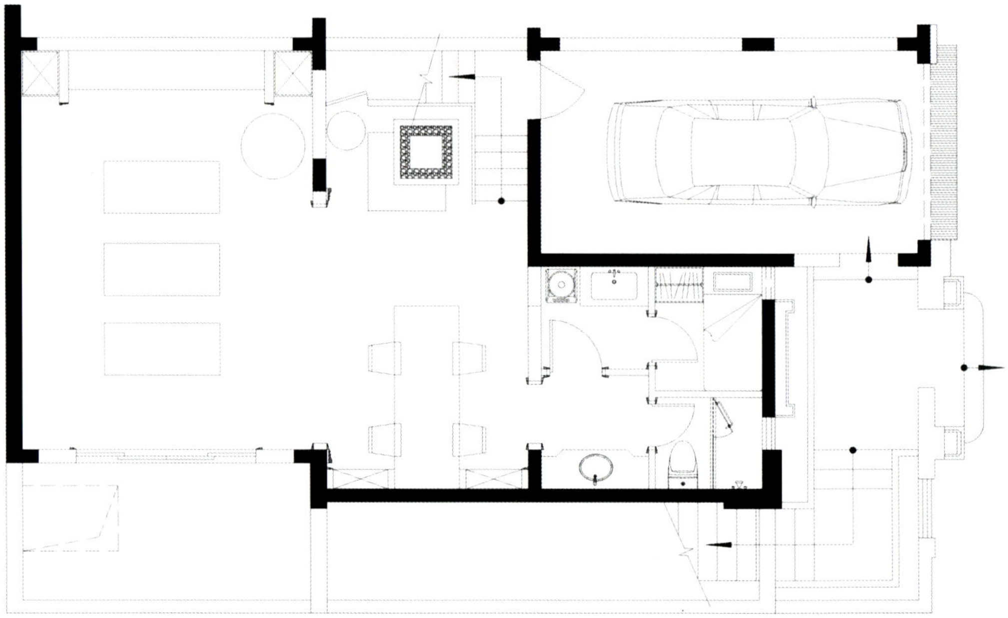

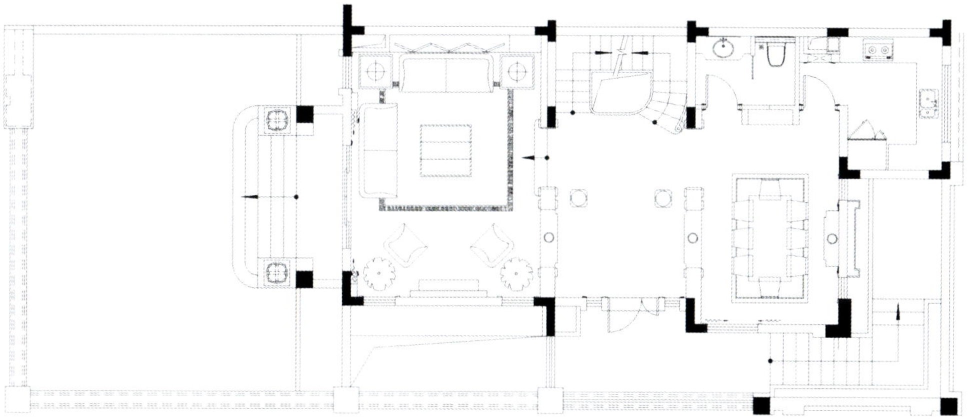

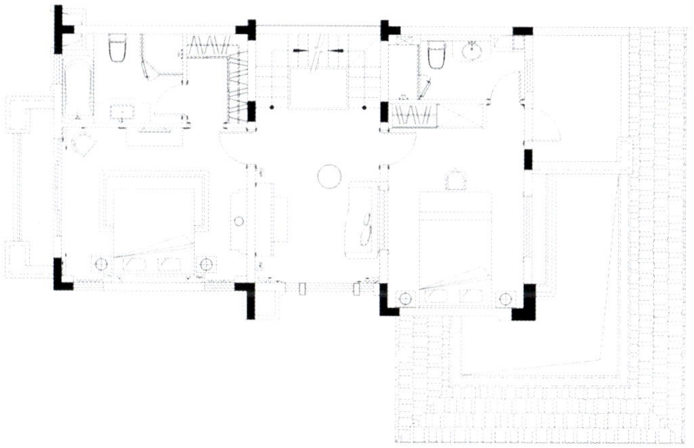

本案设计运用了法式风格，明亮开敞，设计师们以精致典雅和柔美浪漫作为设计的主线，强调了甜蜜浪漫的生活氛围。

本案一层进厅两边增加了立柱式门洞，加强了立面细节和各个功能空间的联系。二层可爱的女儿房布满了各式玩具。白色的线角，柔美的纹饰，展现低调奢华，这一切都将圆女主人一个甜美的公主梦。二层钢琴室就在女儿房的隔壁，能让女儿更好地展现艺术天分。三楼主人房采用柔美的线条、雅致的皮质软包和银箔贴面，充分展现法式的浪漫。曼妙的水晶灯饰使空间更加灵动。地下一层以休闲功能为主，也是女主人展示高超裁剪技艺的地方，女主人的灵感在这里将得到充分释放，充满生活化的场景使人备感温馨。

地点	面积	设计师	设计公司	主要材料
杭州	280平方米	梁苏杭	东易日盛装饰集团杭州分公司	涂料、墙纸、爵士白大理石、樱桃木拼花地板

Neo-classicalism, Hangzhou West Lake

杭州西湖新古典主义

The design of the project is oriented to neo-classical style, integrating the simple style, so the Chinese and Western cultures are perfectly blended. The designer chooses new technology, new materials, and new designing concept to show the internal sentiment to beauty. While paying attention to the decorative effect, the designer uses modern technique and technology to restore the classical temperament which endows the space a dual aesthetic effect of classical and modern. The perfect combination allows people to enjoy the material civilization and mental comfort at the same time.
Regarding the interior materials and colors, the design adopts the Chinese traditional middle, symmetrical, and balanced layout, showing the elegant taste of classical accessories. The design of details shows the delicate and simple designing technique.

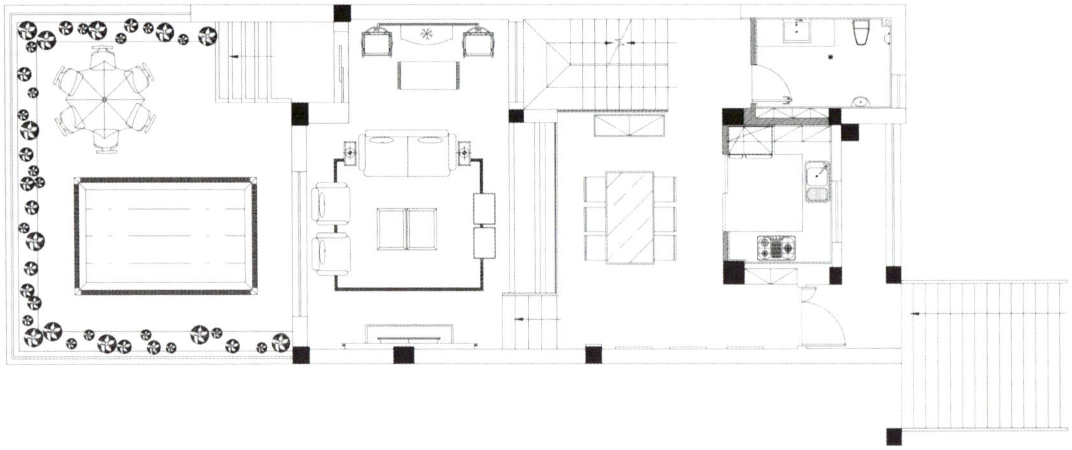

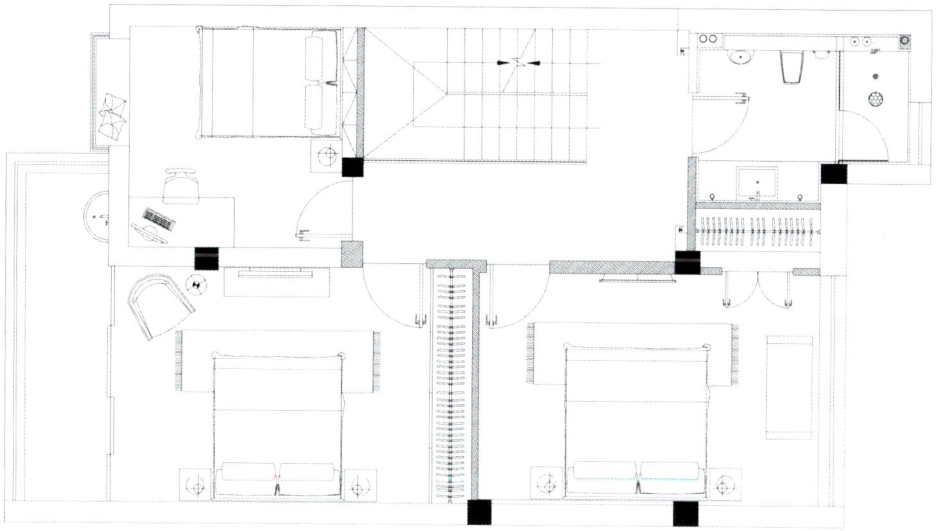

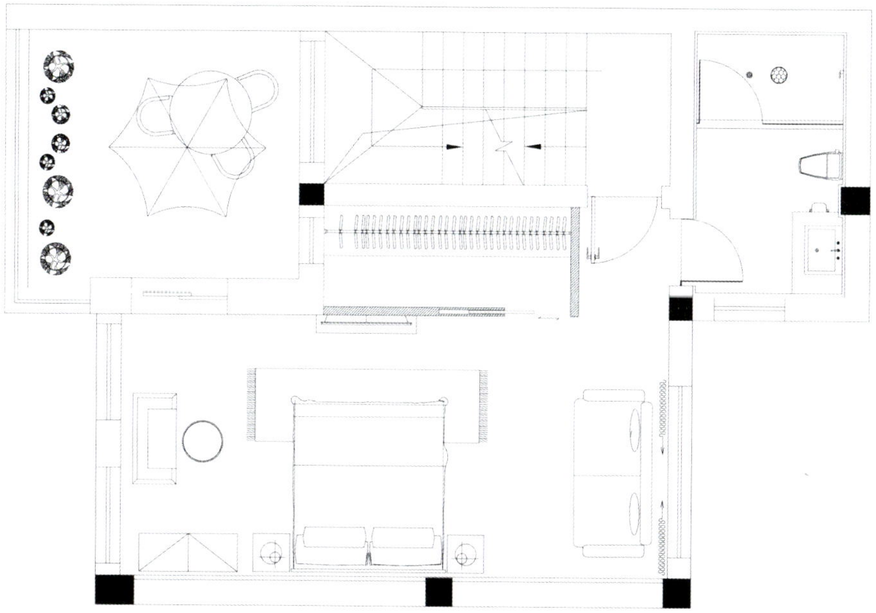

本案的设计定位为新古典主义风格，融合了简约手法，使中西文化在此完美交融。设计师采用了新的工艺、新的材料和新的设计理念去表现内心对美的感悟，在注重装饰效果的同时，用现代的手法和工艺还原了古典气质，使居室空间具备了古典与现代的双重审美效果，其完美的结合也让人们在享受物质文明的同时得到精神上的慰藉。

在室内的材料、色彩方面，设计具备了中式传统居中、对称、均衡的格局特点，又有对古典饰品优雅味道的呈现，细节的设计体现了精致简约的设计手法。

地点	面积	设计师	设计公司	主要材料
上海	380平方米	萧爱彬	萧氏设计	香草米黄大理石、香格里拉地板

The Dream of Mountain She

佘山的梦想

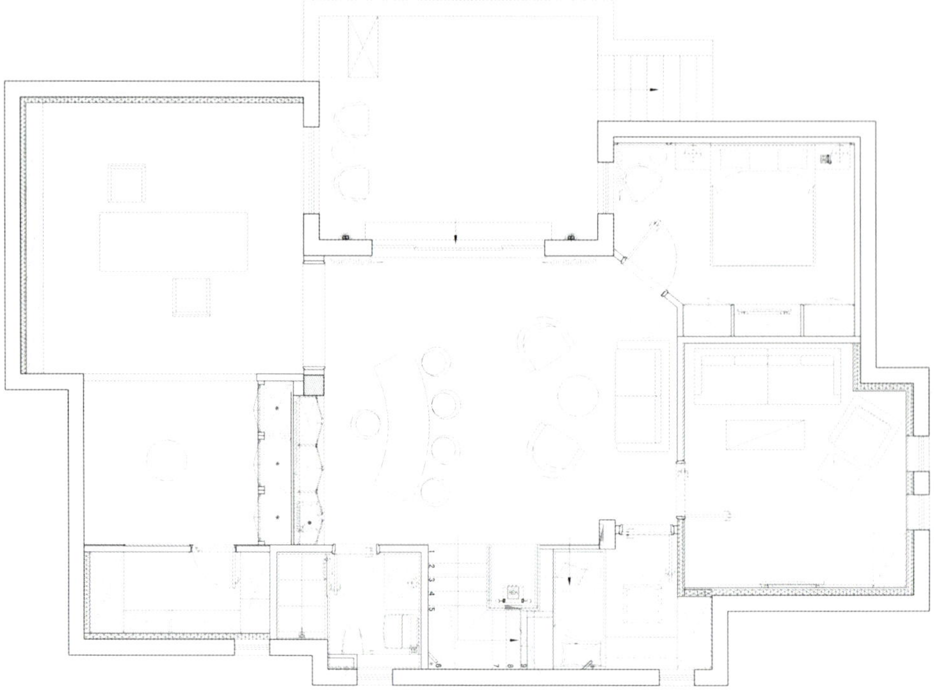

The No. 3 of Mountain She is a pure land in the bustling urban of Shanghai, where many a villas are located in the shadow of bushy greenery. The somewhat Zen space makes every visitor calm down right away. The painting drawn by Xue Song makes one feel the aura of vanguard and fashion. The tall door and the door frames are especially wide. The dining room and living room give one the impression of a garden which the scenes change per step in different space. The designer attaches much importance to spatial psychology, and as a consequence, the elaborated space makes one feel comfortable and pleasant.

The designer considers vision and functions are the most important parts in space followed by materials and workmanship. In this design, the designer presents these specialties in an organic way, so that it creates a comfortable experience in visual effect, and an aesthetic enjoyment in materials and workmanship, in which the mastering of technology and the adoption of materials are just to the point.

佘山三号是上海繁华都市的一块净土，浓密的绿阴掩盖着栋栋别墅，略显禅意的空间令每位参观者都会马上安静下来。入门厅的薛松的画作让人感受到前卫和时尚的气息。高挑的门及门框格外宽阔。西餐厅与起居室使人产生步移景异的园林观感，随着空间产生变换。设计师重视空间心理学，经推敲的空间使人觉得舒适和愉悦。

设计师认为视觉和功能是空间最重要的部分，然后才是材质和工艺。在本案的设计中，设计师把这几种特质有机呈现，在视觉上营造了一种舒心体验，在材质和工艺上产生美感享受，工艺的把握和材质的选用又都恰到好处。

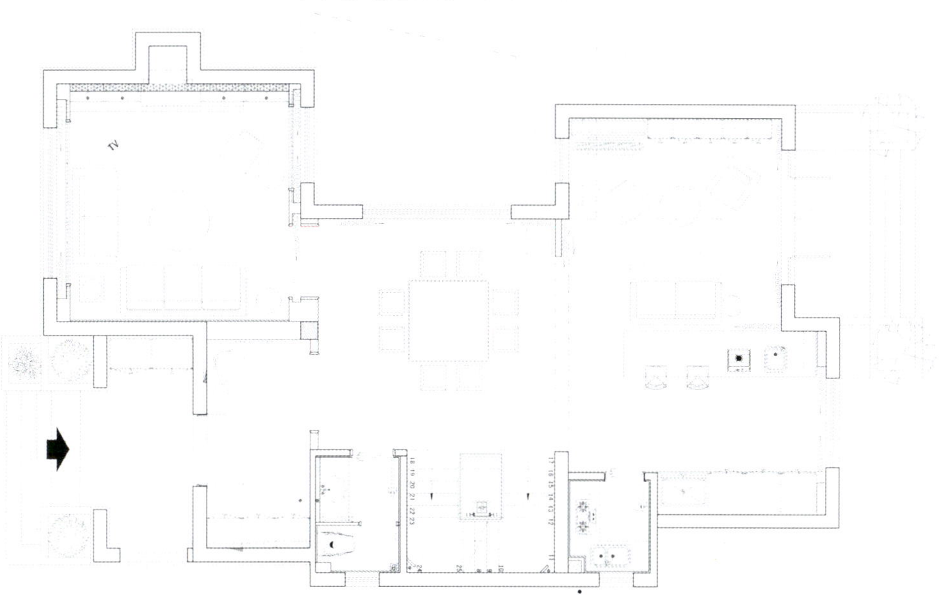

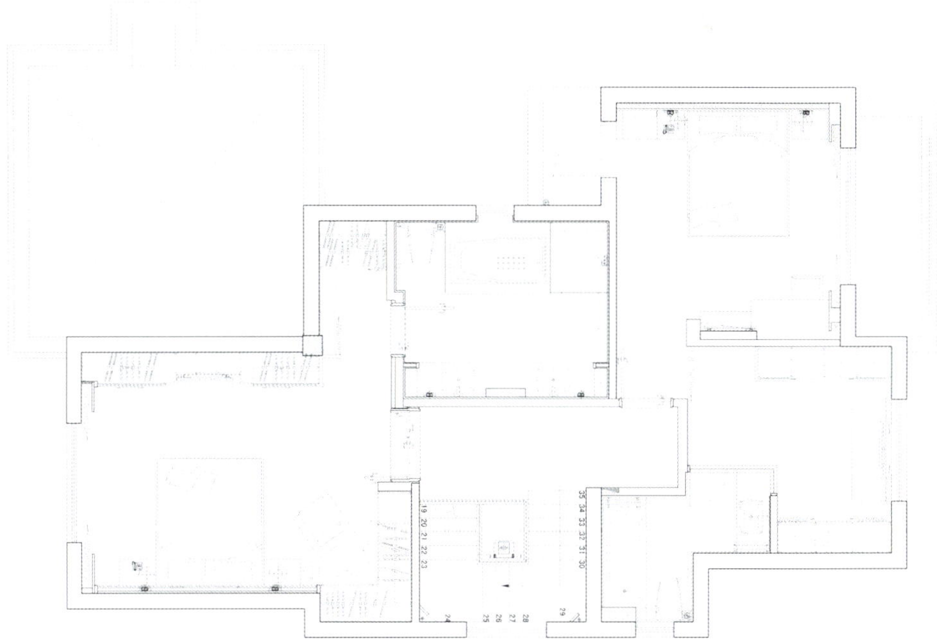

地点	面积	设计师	参与设计	设计公司	主要材料	摄　影
长沙	512平方米	陈志斌	李智勇、谭丽、彭辉	鸿扬集团 陈志斌设计事务所	铁力木、紫彩麻、爵士白、黄洞石、墙纸、陶瓷马赛克	管盼星

Red Sandalwood Palace

紫檀宫

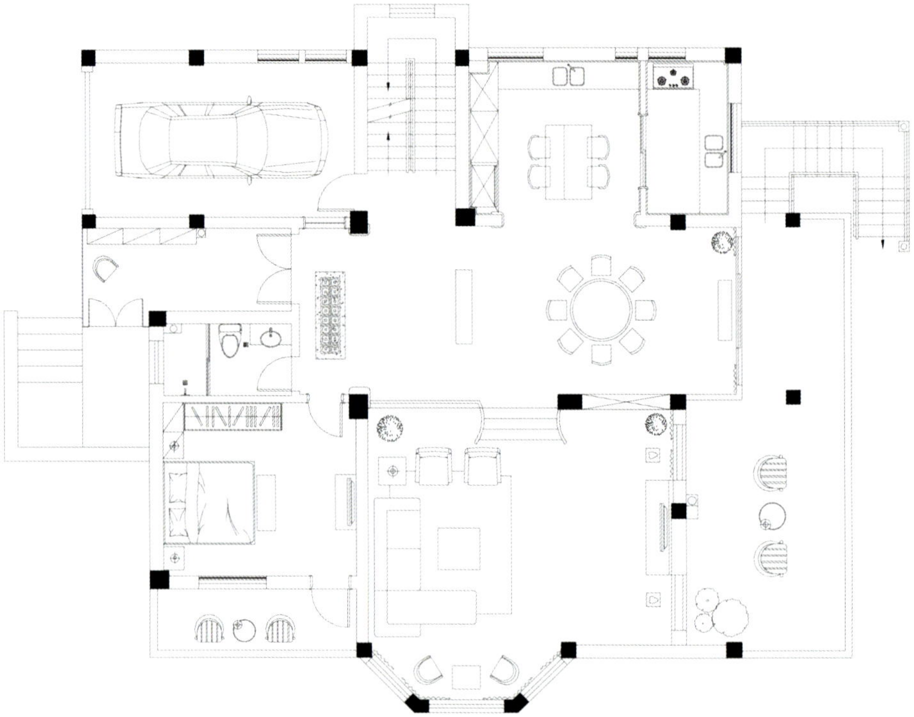

There are many Bubinga, Wenge, Padauk and other solid wood collections in the owner's home, among which Padauk furniture is the most, with the functions of actual use and appreciation, the designer takes this as the theme concept of design.

The living room with the main element of Padauk is simple and generous, and combined with lattice strip screen and other traditional elements, the indoor atmosphere is harmony and low-key. The wallpaper and curtains with similar style echo with each other, and the droplight with traditional patterns of flowers matches the cultural theme of the living room to perfection. The dotted cultural theme makes the Padauk carvings more quite, exuding a unique charm. Inside the house, scene will change at every step, making people linger on.

由于业主家中有许多花梨、鸡翅、紫檀等实木收藏，且其中紫檀木家具最多，兼具实际使用和鉴赏把玩的功能，设计师就以此作为设计的主题概念。

以紫檀木为主要元素的居室空间简约而大度，辅以花格条屏等传统元素，室内氛围和谐而内敛。相似风格的墙纸与窗帘产生呼应，传统花草图案的吊灯与居室内的文化主题搭配得恰到好处。点缀的文化主题使紫檀雕刻更显沉静，散发着独特的魅力。室内步移景异，令人流连。

面积	设计师	设计公司	主要材料	摄影师
216平方米	范义峰、杨俊、黄婷婷	上海天鼓装饰设计有限公司	石材、墙纸、皮质软包、镜面、马赛克	范义峰

Haier Tianjin Villa, Unit A

海尔天津别墅A户型

The design is based on the new Oriental style. Without much flaunt, it applies simple and elegant style. The natural sense of deep color wood structures and sense of days' deposition of contemporary Chinese-style patterns are integrated into the whole design, bringing you a strong humanistic flavor. The floor with wood texture is integrated with the wall into a whole, smooth and rich of texture, bringing you a sense of low-key luxury. The bright droplights are used in each space, making the space lovely and shiny. The solid furniture enhances the sense of steadiness in the bedroom. You can enjoy the scenery outside the window. The sunlight can go through the bedroom easily, where there are virtual water lily and real sunshine. The virtual and real are integrated here which makes the bedroom especially warm and cozy.

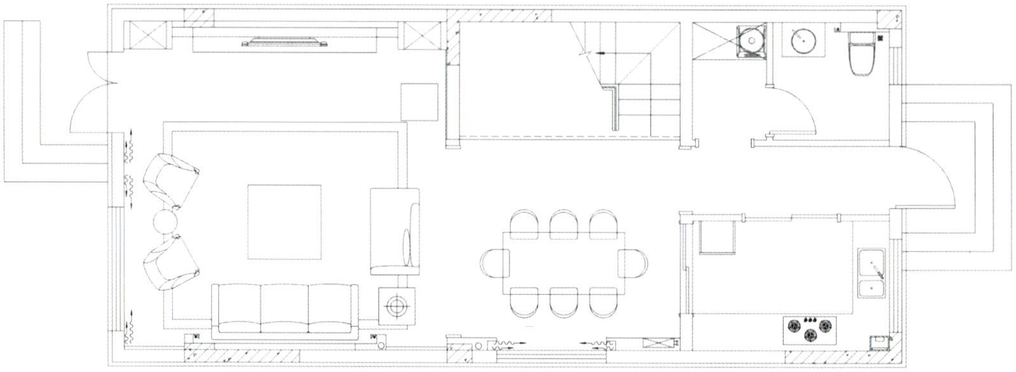

整个设计以新东方风格为主基调，没有太多的张扬，而是采用了简雅的格调。深色木结构的自然感与当代中式图案的岁月积淀感融入到了整个设计当中，给人一种浓浓的人文气息。地面的木纹砖材质与墙身浑然一体，细腻而富有质感，带给人低调的奢华感。明亮的吊灯装饰了每个空间，使空间熠熠生辉。厚实的家具增添了卧室的稳重感。打开窗户便可以看到窗外的景致，阳光也能够轻松地射进卧室，虚的荷花，实的阳光，虚实在这里融合，让卧室变得格外温馨。

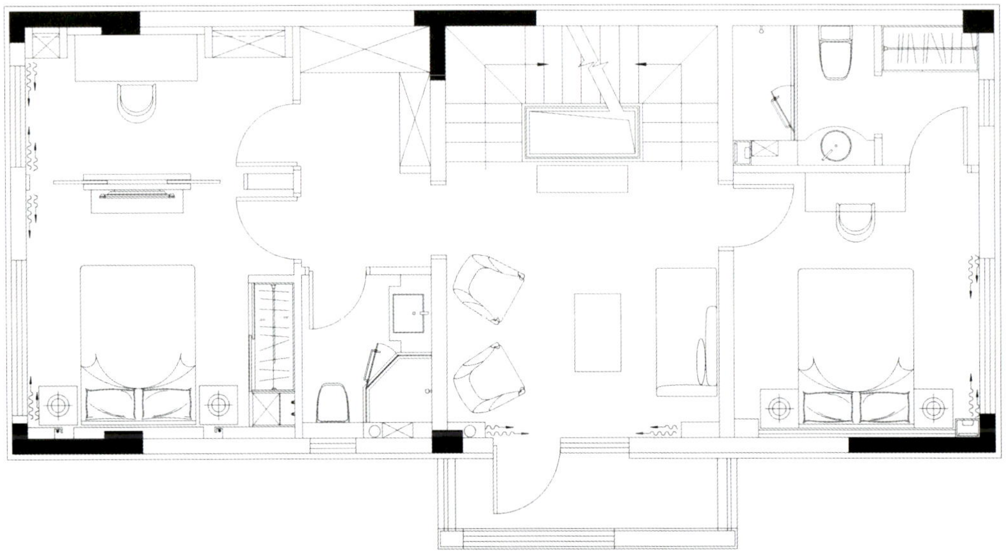

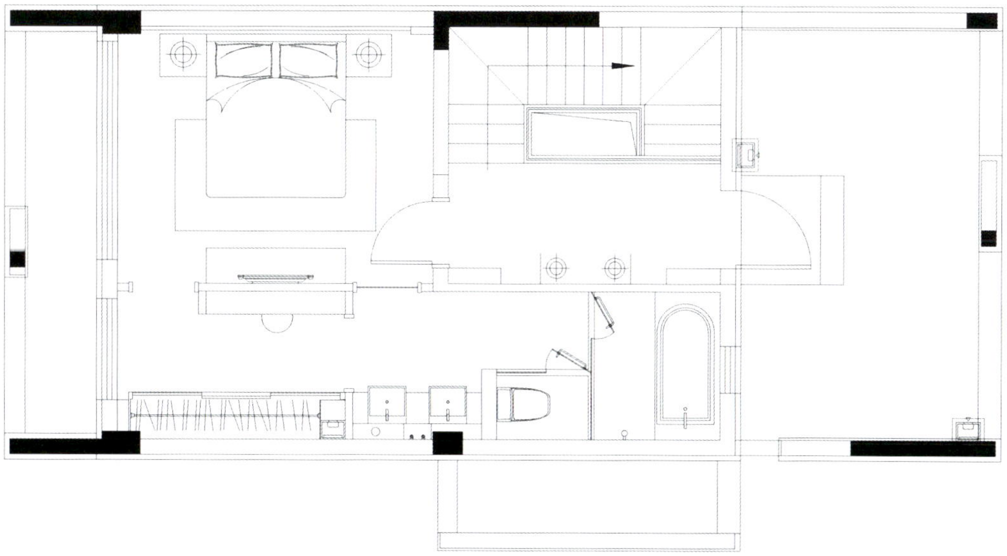

地点	面积	设计师	设计公司	主要材料
哈尔滨	434平方米	万文拓、彭秋波	深圳市品源装饰工程有限公司	艺术挂画、软包、壁纸、大理石、地毯

Swiss Style of "Knight Fashion"

"骑士时尚"瑞士风

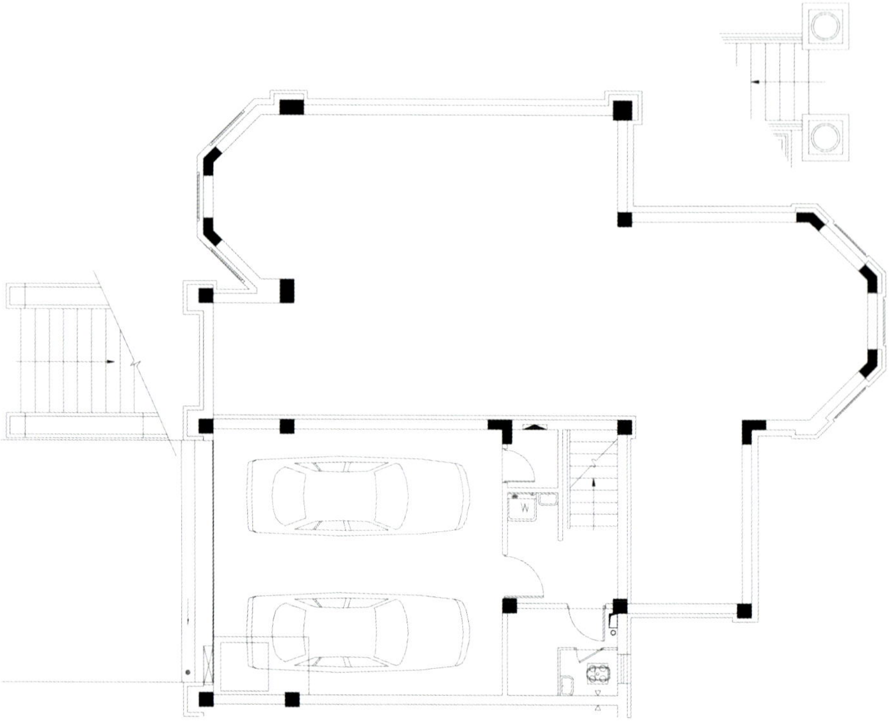

For some time, the minimalism swept the fashion world, and the post minimalism gradually produced fission and began to introduce classic elements which have been abandoned. Today, the "new nostalgia" and "new decoration" greatly favored by the fashion world are interpreted totally, gradually infiltrating into people's lives.

Under the influence of the diversification, the designers do not simply copy the classical, but to integrate the classical into the modern design. The elegant patterns and lines, modern materials and technology, classical atmosphere coupled with modern and elegant furniture, all those are introduced by the way of collisions and conflicts to integrate and show a luxurious sense of fashion. The designers' profound grasp on the materials and shapes, unique perspective on the classical style, grasp and study on the scale, all endow the designing language with fashionable atmosphere, showing the elegant and luxurious temperament unobtrusively one by one.

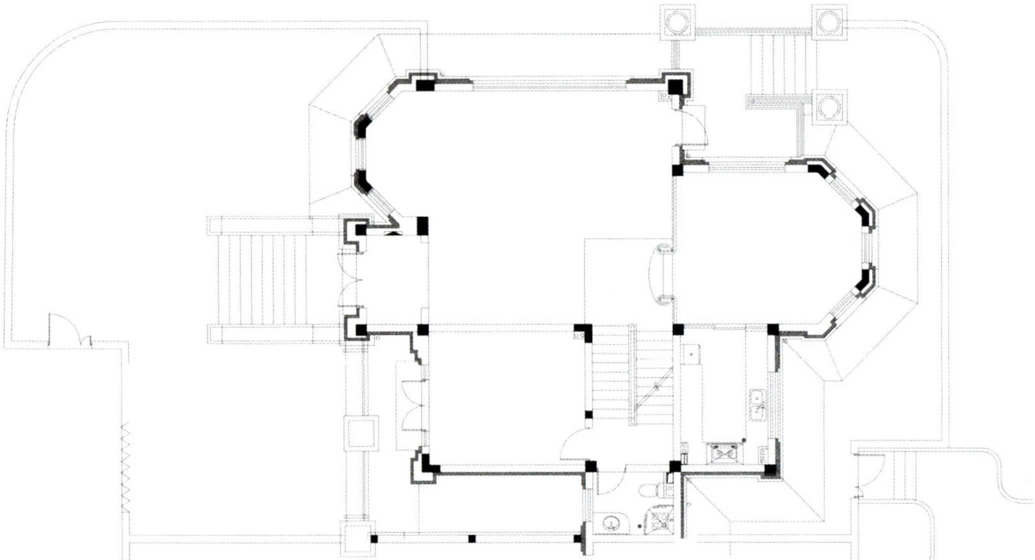

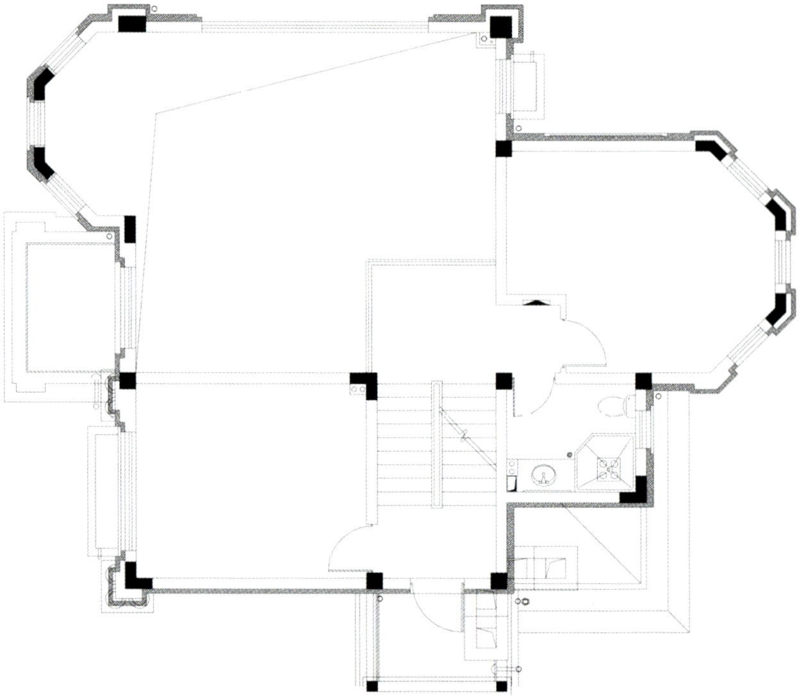

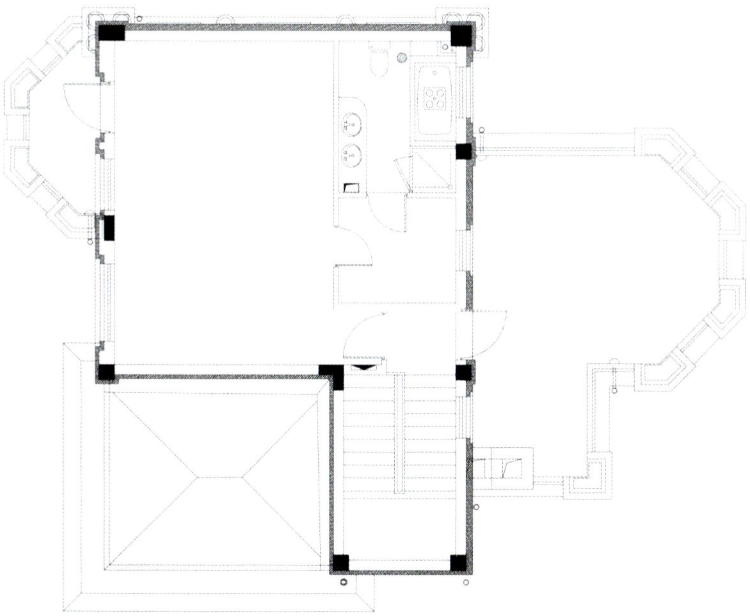

一段时间以来，极简主义风靡时尚界，后极简主义逐渐产生裂变，开始采用那些被摒弃的经典。如今，备受时尚界青睐的“新怀旧”“新装饰”得以全新演绎，也逐渐渗透到人们的生活中。

在多元化的影响下，设计师绝不是简单地复制古典，而是要将古典融入现代中。优雅的图案与纹路，现代的材质与工艺，古典的氛围搭配现代的典雅家具，用碰撞与矛盾的手法，融合出奢华的时尚感。设计师对材质及造型的深刻把握，对古典风格的独特见解，对尺度的掌握和考究，无不使设计语言充满时尚气息，使优雅奢华的气质毫不张扬地一一呈现。

地点	面积	设计师
Rancho Santa Fe, CA	975平方米	Safdie Rabines Architects

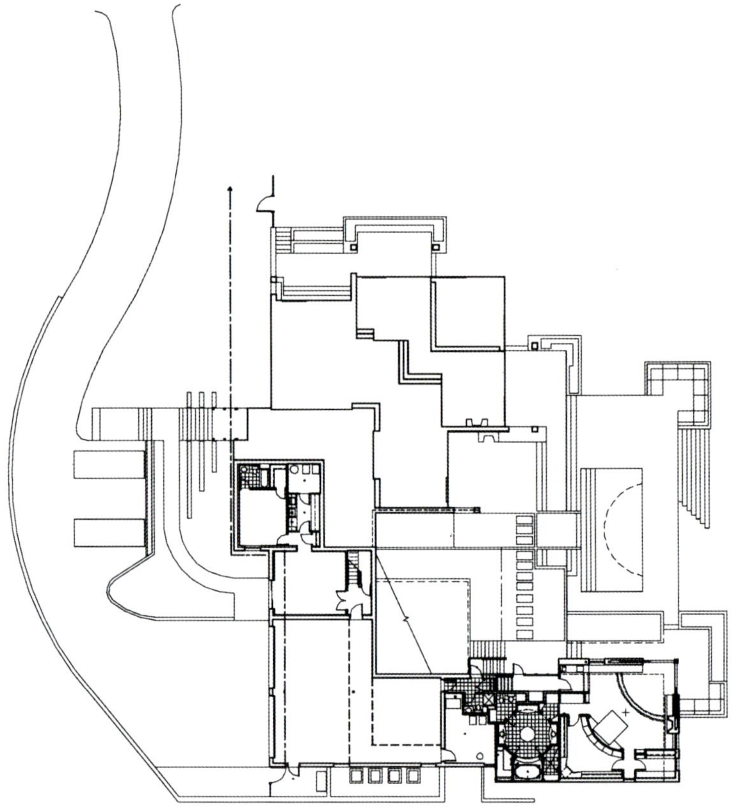

The Hilltop House

山顶住宅

Due to the home's hilltop placement, the building steps gently down the hillside, following the contours of the land. Angled roofs emphasize the stepping quality of the house and provide for sky views, while washing the interior in sunlight and further allowing for a connection with the outdoors.

In order to accentuate these multi-directional views, the home features many wall-sized sliding pocket doors throughout. A distinction between public and private space is created within the house. The more public areas, consisting of several living rooms, sitting areas, kitchen, and dining room, make up the main body of the house, defined by one large space containing smaller "nook" spaces which create a dynamic opportunity for interaction. The portion of the house containing the bedrooms and study are situated around an intimate courtyard area, allowing for additional privacy and solitude.

A large amount of natural wood is used in the house. On the interior, exposed wood ceilings create a warm golden glow throughout the house. Stone flooring is incorporated widely throughout the interior and exterior, often blurring the line between inside and out. The use of these materials adds a sense of warmth and nature to the home while weaving together the integration of space.

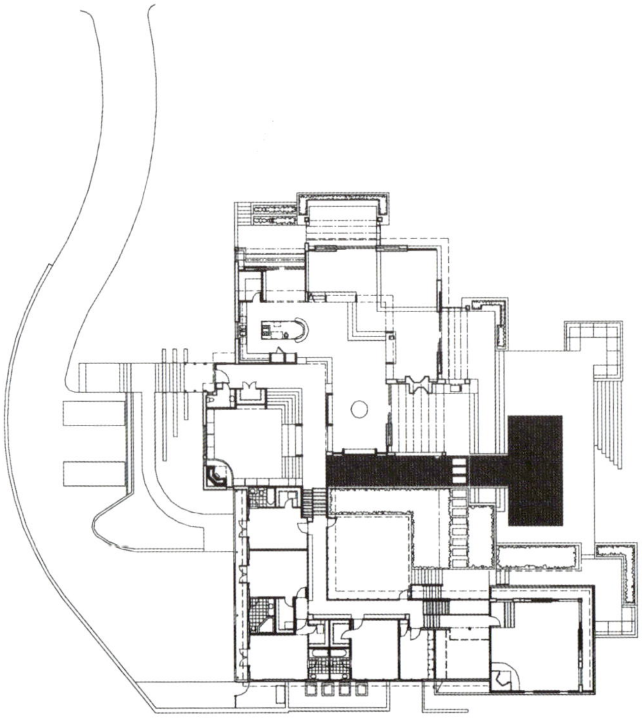

由于房子的位置位于山顶，因此，房屋沿着山坡的地势缓缓地延伸到山腰。棱角分明的屋顶强调了房子的优质质量，并为欣赏窗外景致创造了条件，室内空间因此沐浴在阳光下，并进一步增强了与户外的连接。

为了强调这些多角度的视野，建筑在里外均设计了多道如墙面的拉门。由此将公共空间和私密空间在建筑内部进行了分隔。在较为公共的区域内，包括几间起居室、客厅区、厨房和餐厅，构成了该建筑的主体部分，包含一个个小“角落”空间的大型空间得以界定，这些小“角落”空间创造了动态的互动性。建筑中包含卧室和书房的部分围绕在一处私密的庭院周围，从而提供了更多的私密性和独处的空间。

建筑中应用了大量的自然木材。在室内空间中，无遮无掩的木天花形成了一种温馨的金色光晕萦绕着整个建筑。室内外铺设了大量的石地砖，模糊了室内外的界线。这些材料的运用不仅将各个空间有机地衔接起来，更为该项目增加了温馨自然的感觉。

地点	设计公司	摄影师
加拿大	Linebox Studio	Christian Lalonde

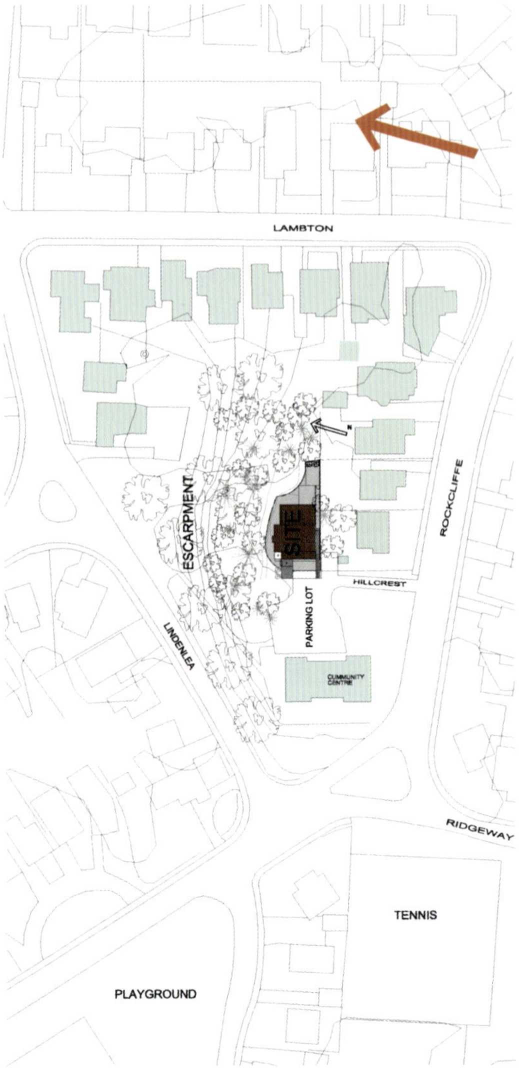

Lighthouse

灯塔

Lighthouse was designed for a family of five, looking to relocate from a rural setting to a modern urban dream house. The design is based on a "vertical stacking of both open and closed spaces and minimal details that establish an ascending hierarchy of social and private spaces." Splayed around a spiraling steel and maple staircase that rises within a vertical glass shaft are rooms that exude airiness and a sense of connectivity. "The large side entrance with a limestone floor which leads into the open kitchen and dining area, creates a generous 'shedding' area" for the teenage kids. The sleek Irpinia cabinetry is "neutral in tone and is spatially defined by a large granite-topped island with a lower, table-height extension" for spreading out homework or gathering around to socialize. The space is characterized by views of exterior foliage and Mooi pendant lights that bring down the sense of scale and soften the otherwise crisp, rectilinear volumes.

The elevated great room with its 22-foot high ceilings, warm maple floors (all floor have radiant heating), and crisp white walls is uncluttered and absent of visible light fixtures. This room showcases art and facilitates conversation. Floor-to-ceiling commercial-grade windows wrap the north corner of the room, expanding the space into the forest beyond. The two-sided fireplace that screens the front entrance provides a "welcome beacon of warmth to new arrivals on a cold night" and fills the living room with its playful glow. The second storey has a sizeable landing that wraps around the staircase shaft and serves as the children's rooms, a bathroom and a loft office overlooking the double-height living room. The top "tree house" level is a boutique hotel style master suite and roof-top terrace with spa and spaces for working, reading and sleeping.

WARHOL LIVE
HERB RITTS

“灯塔”是专为五口之家设计的，是他们从乡村迁移到现代都市的梦之家。本案基于“开放式和封闭式空间的垂直罗列和建立公用和私密空间等级层次的最小细节”而设计。螺旋的钢质械木楼梯及垂直的玻璃围成几个空间，具有良好的通风和连通性。“大型的侧入口配上石灰岩地板，为十几岁的孩子营造出大气的‘撒欢’区，引导人们进入开放式厨房和用餐区。”光滑的伊尔皮尼亚橱柜采用“中性色调，空间上由一个花岗岩铺面的‘小岛’及稍低一些与桌面齐高的延伸部分界定”，人们可以在此整理家务或聚集在一起进行社交活动。空间以室外的绿色植物及室内的Mool吊灯为特点，降低了规模感，柔化了明快的直线造型。

挑高的大房间设有6.7米高的天花板、温和的枫叶地板（所有地板都设有辐射采暖），明快的白色墙壁整洁且没有明显的灯源装饰。房间充分展示了艺术与便利的完美结合。商务级落地窗户占据了房间的北角，将空间扩展到室外森林。双面壁炉遮盖了前面的入口处，“在寒冷的夜里为来访者点亮温暖的迎接灯塔”，使客厅充满活泼快乐的光辉。在二楼，一块相当大的空地围绕着楼梯通道，设置有儿童房、浴室和阁楼办公室，可俯瞰下面的跃层客厅。最上面的“树屋”层是精品酒店风格的主人套房，屋顶露台设有水疗区和工作区，人们可以在此读书或休息。

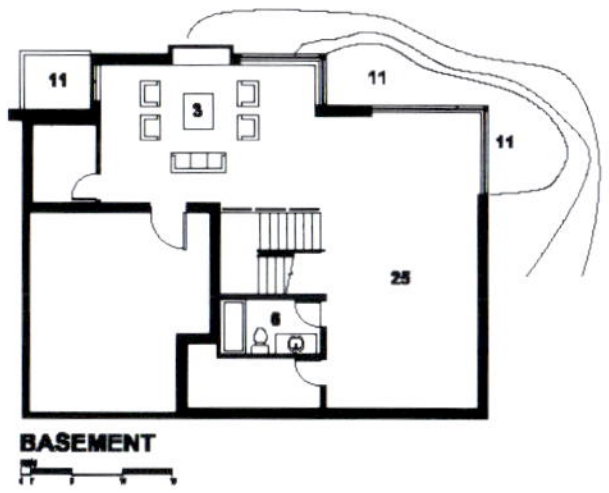

BASEMENT

LEGEND

1. FRONT ENTRANCE
2. FRONT LOBBY
3. LIVING SPACE
4. DINING SPACE
5. EATING AREA
6. SIDE ENTRANCE
7. MUD ROOM
8. POWDER
9. SINGLE CAR GARAGE
10. PLATFORM DECK
11. SUNK-IN GARDEN
12. OPEN TO BELOW
13. BEDROOM
14. BEDROOM
15. WASHROOM
16. LAUNDRY
17. BEDROOM
18. OPEN OFFICE
18. SOCIAL SPACE
20. WALK-IN CLOSET
21. ENSUITE
22. MASTER BEDROOM
23. ROOF TOP PATIO
24. DOG RUN
25. GAMES

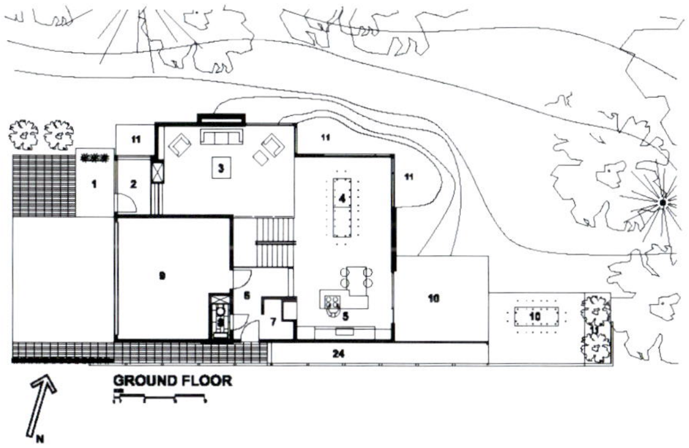

GROUND FLOOR

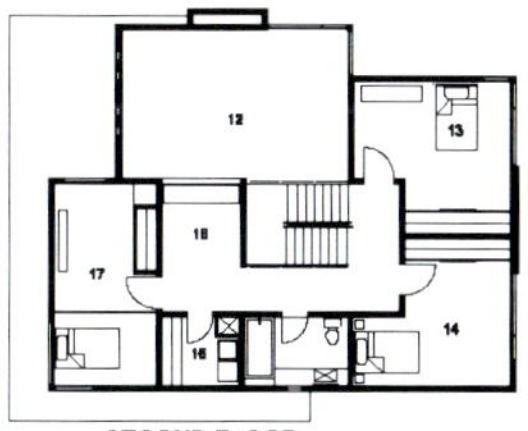

SECOND FLOOR

THIRD FLOOR

设计师

Nacho Polo

ENCARNACION
恩卡那森

Mirror table is a Nacho Polo design for this home. The kitchen is a Nacho Polo design for this home also and the material is Corian & Wood. The chandelier is from an antique market from London S XIX.
The corridor begins in the foyer of the house and ends in the master bedroom, in the middle there is a guest room. The white frames are made of wood that bought in different countries around the world. The designer wanted to create a baroque but fresh and contemporary atmosphere. The dressing room is located in the master bedroom suite, where you also can find the bedroom, the bathroom and a studio.
It was designed in particular for this home and later was produced by the name of "Twenty-Seven MIRROR" as a limited edition by Nacho Polo WALLPAPER by Designers Guild Handmade black LAMPS.
The wallpaper of the small bathroom is from London and the two pictures are the copy in a different scale of two Modigliani. Small black table and the mirror in the guest bedroom are from antique market MADRID that was painted in black paint outside and red paint inside.

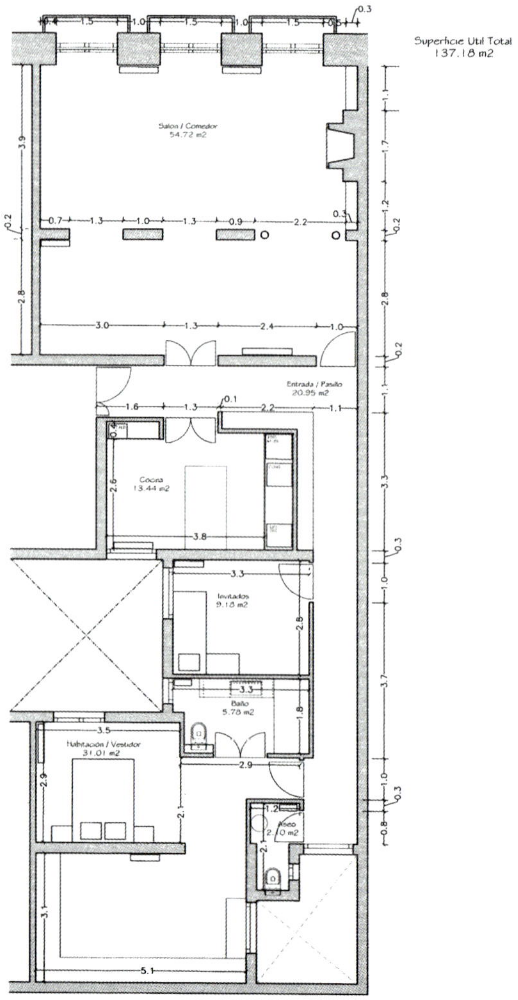

JAPANESE DESIGN

DINASTIA

镜桌是设计师乔纳·波罗针对该房屋设计的。厨房也是乔纳·波罗针对该房屋设计的，原料采用了可丽耐人造大理石加木材。枝形吊灯是从伦敦S XIX古董市场买来的。

走廊从房子的玄关延伸到主卧室，走廊的中间有一间客房。白色的边框是用在世界各地买来的不同木材制作的。设计师意欲打造的是一个颇具巴洛克风格，但又清新现代的氛围。更衣室设在主卧室套房之内，套房中你还会发现卧室、卫生间和书房。

其中的黑色灯具本是专门为该项目设计的，后来以“二十七镜”这一名字得以限量生产，由设计师协会的乔纳·波罗壁纸公司生产。

小卫生间的壁纸购于伦敦，两张画是对两幅莫迪利亚尼作品的不同程度的模仿。客房内的小黑桌子和镜子购于马德里的古董市场，外面漆成了黑色，里面漆成了红色。

地点	面积	设计师	设计公司	主要材料
高要	310平方米	何永明	广州道胜设计有限公司	泰柚直纹木、大理石、马赛克

Park Island Villa

珀丽湾别墅

In order to make the sculpted works further integrating into the living space of the owner, the designer chooses modern simplified European style in white keynote with soft lines, colors and decorations, and makes the stretching force of the sculpture as the focus of the space. When you enter the foyer, you will find that the whole public space has a visual extensibility and adequate lighting, which forms a special and interesting dialogue with the outdoor landscape and show a graceful attitude as well. The objects and scenes in the space are added to different designing elements by the rich and skilled matching experience of the designer. And all these bring the focus of the space back to human. As for life and beauty, they should be cared about most in the space.

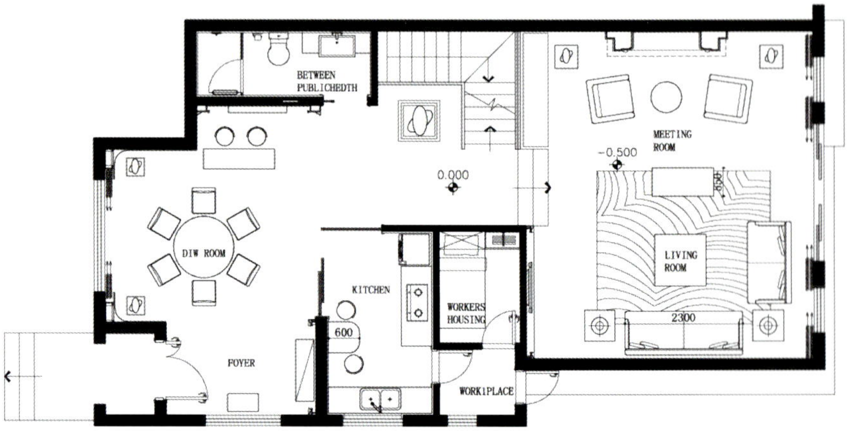

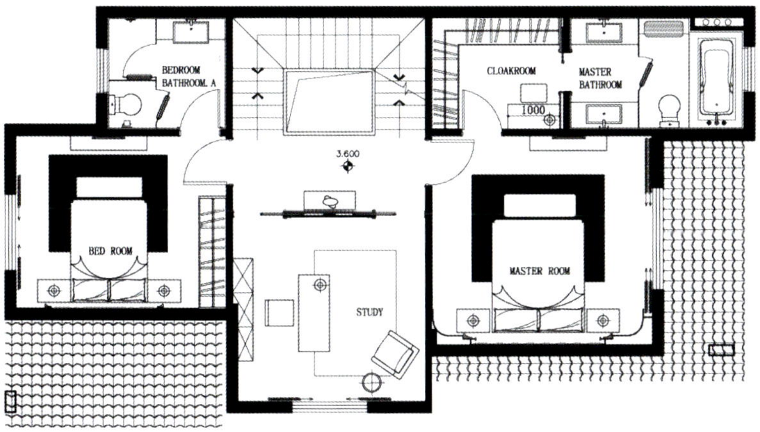
BEDROOM
BATHROOM. A
CLOAKROOM
MASTER
BATHROOM
1000
3.600
BED ROOM
MASTER ROOM
STUDY

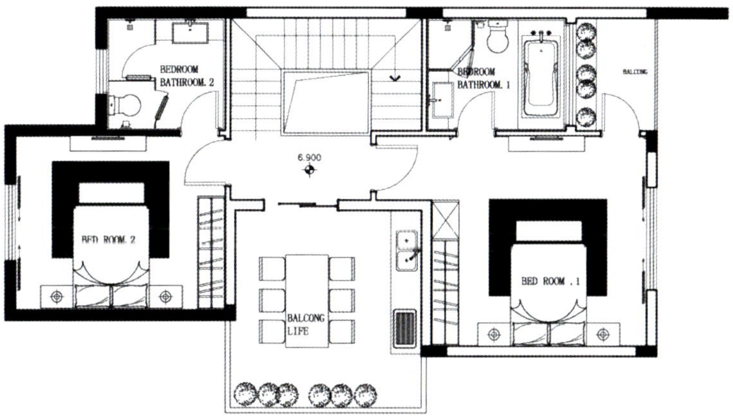
BEDROOM
BATHROOM. 2
BEDROOM
BATHROOM. 1
6.900
BED ROOM. 2
BED ROOM. 1
BALCONG
LIFE

为了使雕塑作品进一步融入屋主的生活空间中，设计师以白色基调的现代简欧风格，淡化空间的线条、色彩与装饰，令雕塑本身的张力成为空间的重点。

进入玄关后，整个公共区域具有视觉延伸性，采光充足，与户外景观形成一种独特有趣的对话，并展现出一种优雅的姿态。空间里的一物一景都被添进了不同的设计元素，在设计师丰富熟练的搭配经验下，彼此烘托、唱和。而这一切，又让空间重新回到人本身。关于生活，关于美，才是空间里最该被在乎的一切。

地点	面积	设计公司	建筑商	摄影师
旧金山	288平方米	Feldman Architecture with Joe Oroza Design	Lara Construction	Paul Dyer

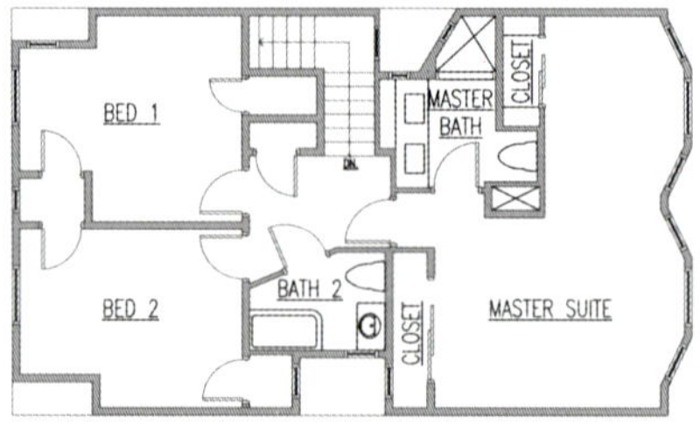

Buena Vista

布埃纳·维斯塔

The Buena Vista renovation began as an effort to transform a segmented two-storey residence into a vertically integrated, contemporary garden home.

By reclaiming the unfinished basement space and additional area under the deck, a new living area emerged with a direct connection to a modern garden. The kitchen was dramatically opened up to the dining room and living room to create better circulation through the residence. The redesign included removal of the fireplace that separated the kitchen from the dining room and replaced one of the dividing walls with channel glass which allows a more direct connection between the spaces.

Oversized pivot doors on both of the main living levels invite inhabitants to enjoy the outdoor decks and dramatic views. The material palette subtly combines wainscoting, warm wood floors and cabinets with steel and channel glass for a contemporary, modern intervention. The chocolate brown cabinets in the kitchen are accented by a few upper cabinets in white and a white Calacatta de Oro marble countertop. As the clients collect contemporary art as well as antiques, a color palette was chosen which blends bold, dark color with light trim and countertops.

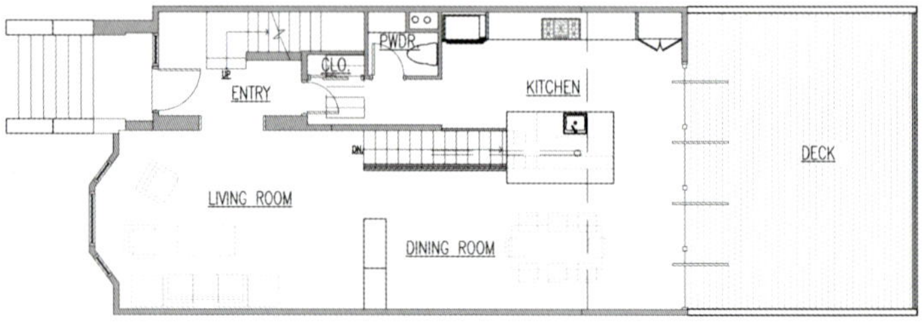

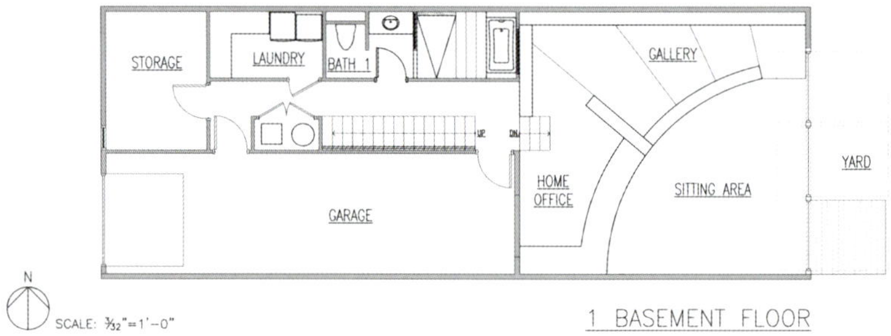

布埃纳·维斯塔方案最初试图将分割为两层的住宅改造成一个垂直整合的现代花园式住宅。

通过重新利用未完成的地下室空间和层面下的附加区域，一个新的生活区展现在人们面前，与现代式花园直接相连。引人注目的是厨房与餐厅和客厅之间完全敞开，营造出居住空间更好的通风条件。重新设计的部分包括拆除分隔厨房和餐厅的壁炉，并利用槽型玻璃取代分隔墙的一面，使空间之间产生更多的直接连接。

主要生活层两侧的超大型枢轴门使居住者可以享受户外阳台和动人美景。材料的选择巧妙地结合了护墙板、温和的木质地板以及当代风格的钢质槽型玻璃陈列柜。厨房里，上层几个白色橱柜和白色花纹大理石台面突出强调了巧克力棕色的橱柜。由于业主喜欢收集当代艺术品和古董，设计师采用了一系列不同的色彩，以大胆的深色搭配明亮的装饰和台面。

地点	面积	设计师	设计公司	主要材料
深圳	400平方米	王冠	深圳市矩阵室内装饰设计有限公司	木纹洞石、红色玻璃、金箔、柚木地板、墙纸

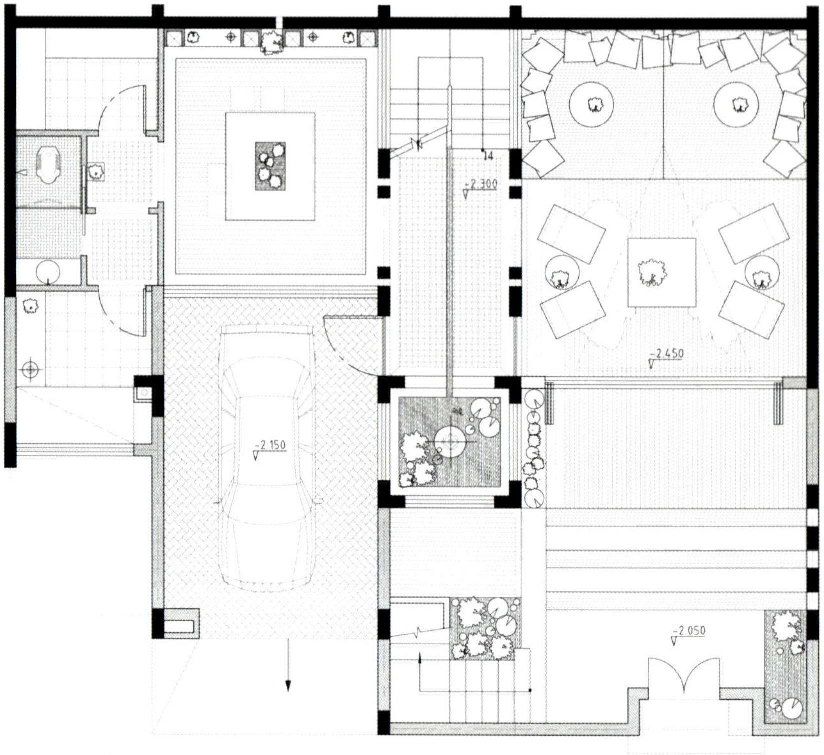

Central Joy City Villa

中央悦城别墅

The biggest characteristic of this project lies in the rich colors and the diversity of the lineal sculptures. Through the adoption of colors, such as the white of the sofa in the living room, the black of the lamp cover, the bronze yellow of the tea table furnishings, the dark green of the cushions, the dark red of the flooring, the light red of the window gauze, the designer decorates the living room with brilliant colors. In addition to that, the bold appliance of red also gives prominence to the originality of the designer. The red elements, such as the red window curtains and the red wall decorations, have developed the passion of red into its full.

The sketching and planning of lines reveal the unordinary skills of the designer. In the living room, the edge lines of the platform and the rectangular tea table form a shape of "回", showing a smooth and harmonious spatial atmosphere. Well in the stairwells, the wave lines formed by the intersection of the walls and the base are like sea waves that are winding ahead. In the space that is composed mostly by straight lines, it reveals, in all respects, the profound thoughts and straightforward character of the designer.

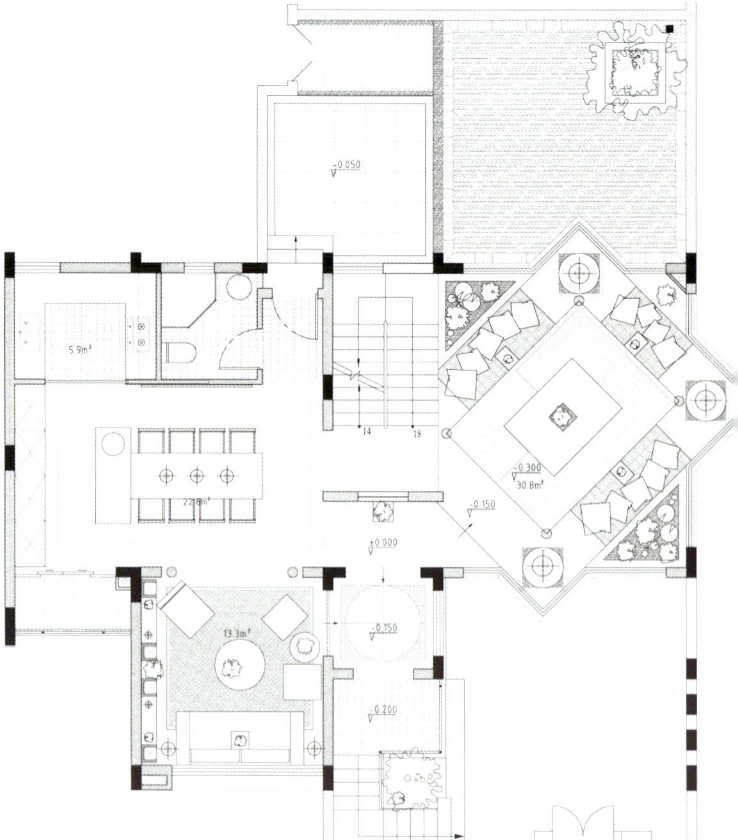
-0.050
5.9m²
14
18
-0.300
30.8m²
22.8m²
-0.150
±0.000
13.3m²
-0.150
-0.200

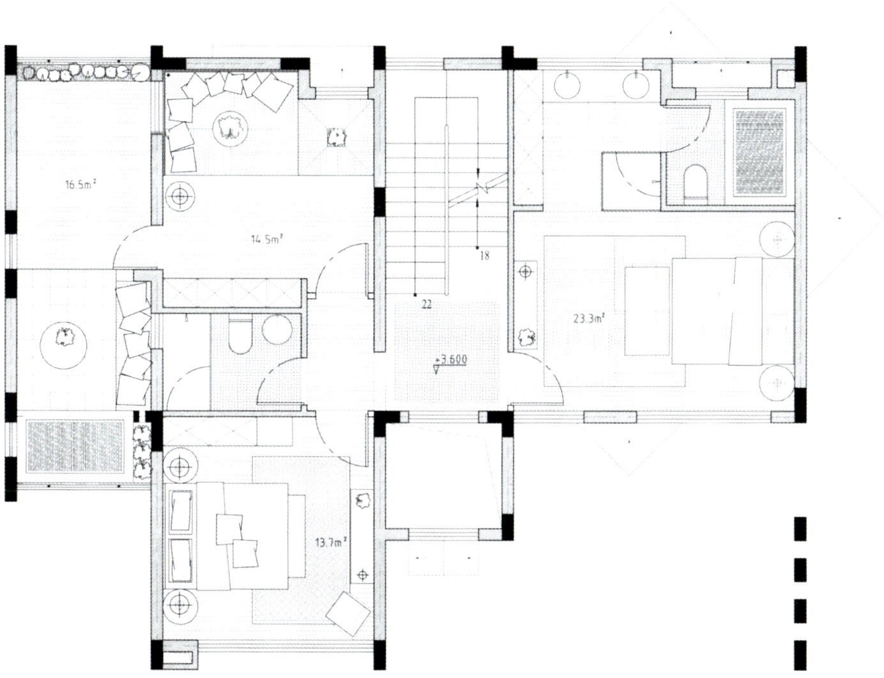
16.5m²
14.5m²
23.3m²
13.7m²
18
22
3.600

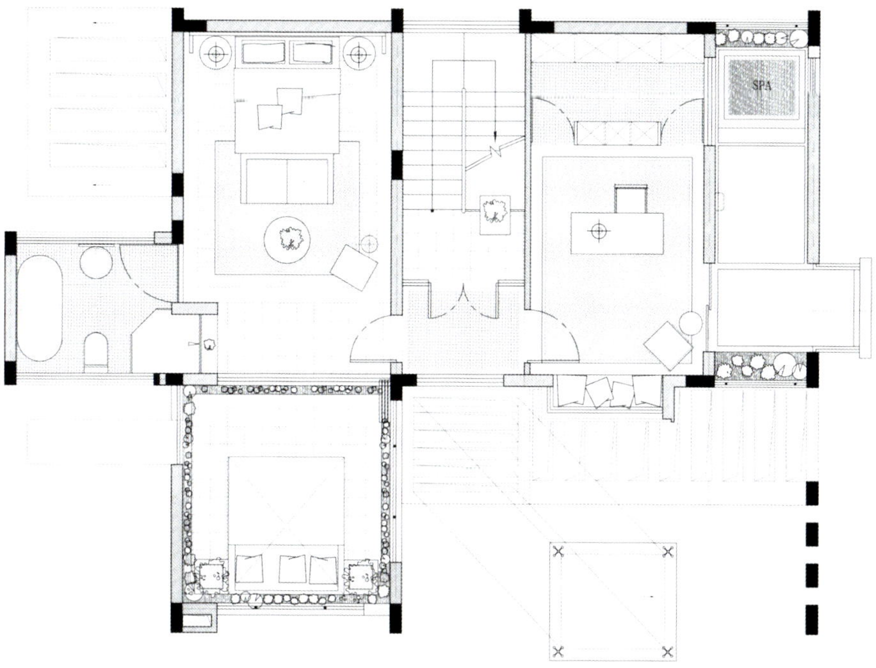
SPA

本案最大的特点在于丰富的色彩和多变的线条造型。设计师通过客厅里沙发的白、台灯罩的黑、茶几饰品的铜黄、抱枕的墨绿、地板的深红、窗纱的浅红等色彩的运用，将整个客厅渲染得五彩斑斓。此外，红色的大胆运用也突出地表现了设计者的别具匠心。红色的窗幔和红色的墙壁等元素，将红色的热情发挥到了极致。

线条的勾勒与构图体现了设计师不凡的功力。客厅里的地台边沿线与方形的茶几围成“回”字形，展现出一个圆融的空间氛围。楼梯间的墙面与底座交汇处形成的曲线如波纹般向前蜿蜒延伸。在直线条居多的空间里，处处展现着设计师深邃的思想与直率的性格。

地点	设计师	设计公司
美国	MARYANN THOMPSON, FAIA	MARYANN THOMPSON ARCHITECTS

Stonington

斯通宁顿

The approach to the house from the west reveals a scheme of three distinct volumes, organized axially along the ridgeline and nestled into a working farm. Existing stone walls and foundations and massive mature Copper Beech trees define the building site. The arrangement of the volumes along the site is a direct response to the geometry of these traces of the property's history and tradition. The west-facing public edge is more insular, while the east and south-facing edges give way to the site and sun through broad expanses of glass and sliding doors that connect the house directly to the landscape. The undulating footprint defines a series of outdoor rooms that address unique views from or through the house to the property, forging connections between the new residence and the existing pastures and nearby outbuildings. Program spaces are distributed between the volumes, creating varied layers of occupation which are altered and adapted through the manipulation of shifting interior partitions. Movement through the house from west to east reinforces ridge as a layer that defines levels of privacy and intimacy, the experience of which culminates in views of the farm and the ocean in the distance from a private balcony, perched atop the living room, in the master bedroom.

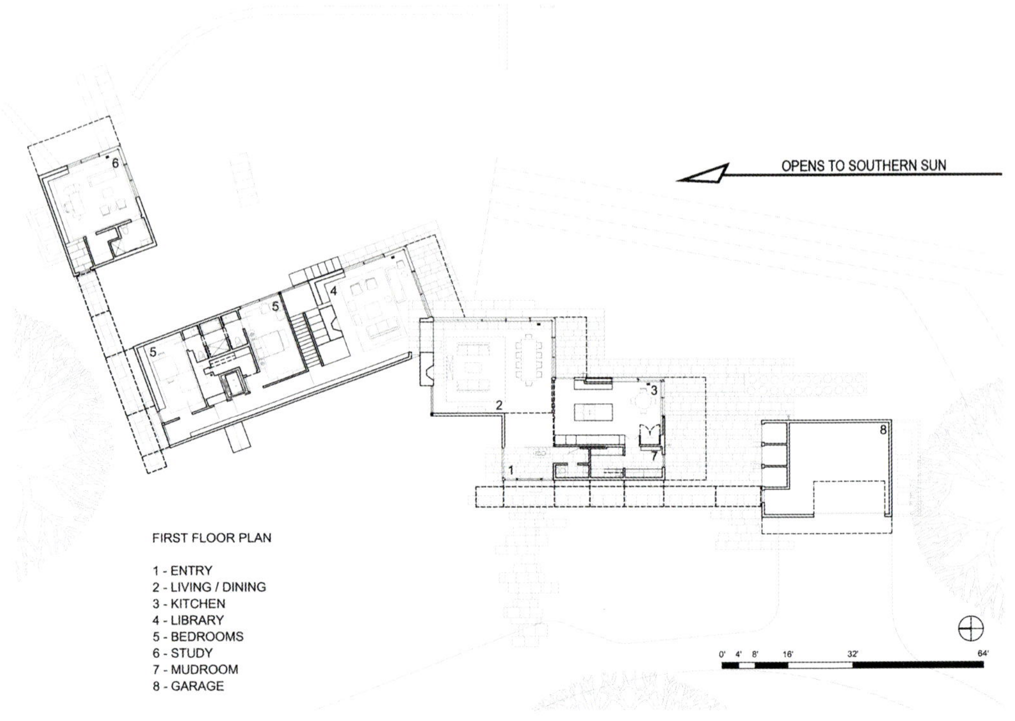

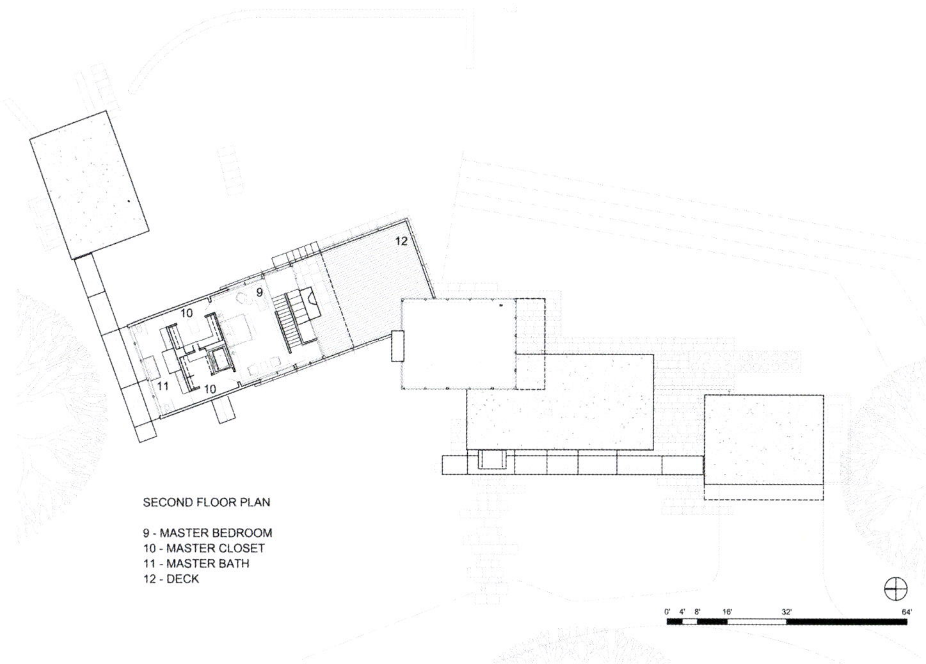

从西面观察，建筑采用了三种不同的体量设计，沿着山脊线的轴向隐藏在一座农场中。原有的石墙、地基和大片的成熟山毛榉树确定了建筑的位置。建筑中的体量设置是对该项目历史和传统几何式的直接呼应。朝西的公共边界更为保守，而朝东和朝南的边界留有空隙，阳光穿过草坪和连接着房屋的推拉门，直接照射到如画美景之上。纵横的脚印表明了一系列的室外空间，强调了室内外的独特视野，形成了新建住宅和已有牧场及附属房屋的联系。项目空间在不同的体量中展开，根据对室内分区转换的处理形成了不同的层级。整个项目自东向西强化了山脊线作为界定隐私和私密程度的划分，最后来到的是位于起居室上面的主卧房里的一个私家露台，在此可以欣赏到农场和远处的海洋。

THE ROAD

面积	设计师	设计公司	主要材料
172平方米	张馨	京轩空间设计/张馨室内事务所	钢刷木皮、清玻璃、天然木皮喷白、古典线板、实木、镜面

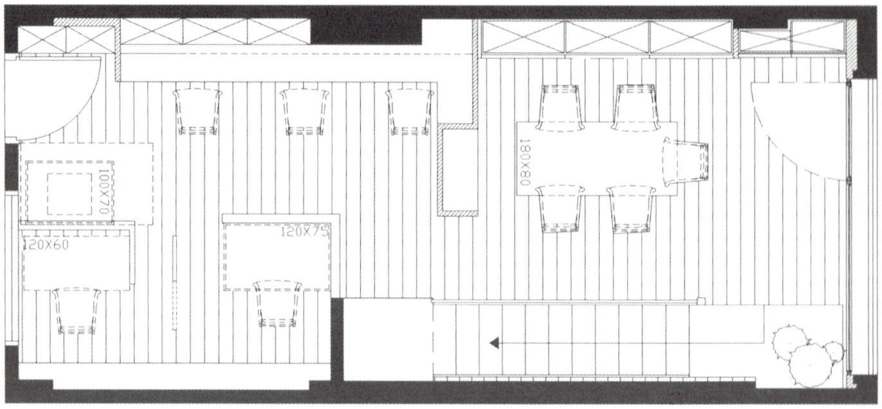

Perfect Integration

完美整合

This building is a detached villa, which integrates the office space and residence space, making the two connected yet independent as well. The first floor is the office space, and the kitchen and dining room are kept for meetings and dinner parties for employees, where the elegant pure white rustic style is adopted, with exquisite and delicate details. The stairs partitions living space into two, separating different areas for study room and bedroom. The dressing mirror in the long and narrow corridor enlarges the spatial dimension, and the ceiling that is without obsessive decorations makes one's visual focus be placed on the graceful French style boards, which highlights the classical sense. The forth floor is divided into a boy room and a girl room, the same size being planned with effects of different styles. In the high ceiling rest room on the top floor, the television background wall and the cabinet matches very well on proportion. It takes gray and white as the key tone, and supplemented with the royal blue colored landing curtains and the strikingly red lamps, taking on an aura of harmony and exquisiteness, appealing the French style classicism in a fantastic way.

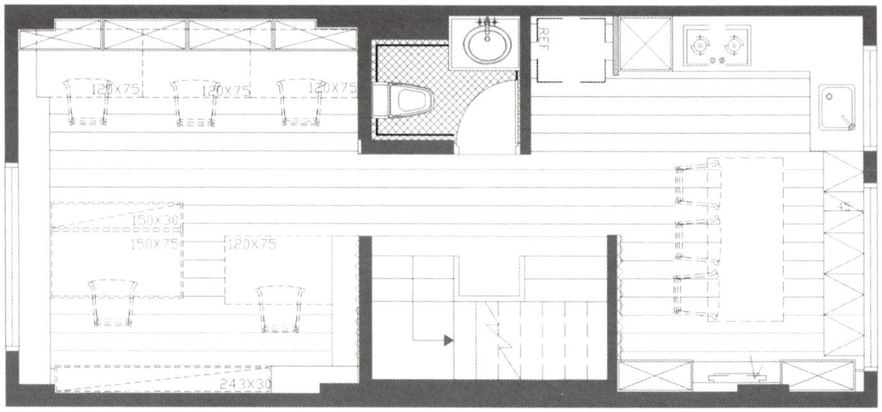

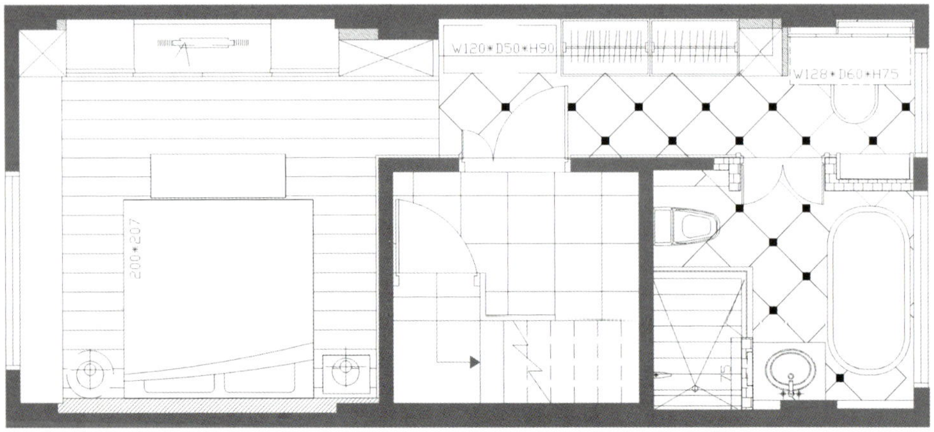
W120*D50*H90
W128*D60*H75
200*207

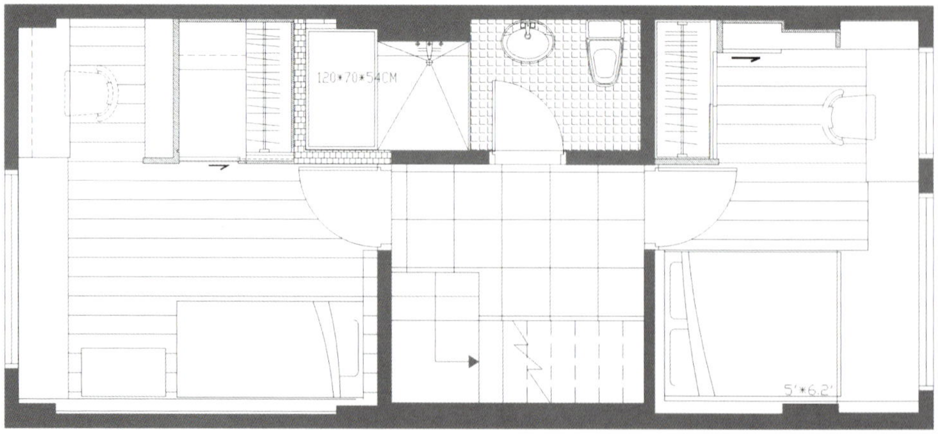
120*70*54CM
5'*6.2'

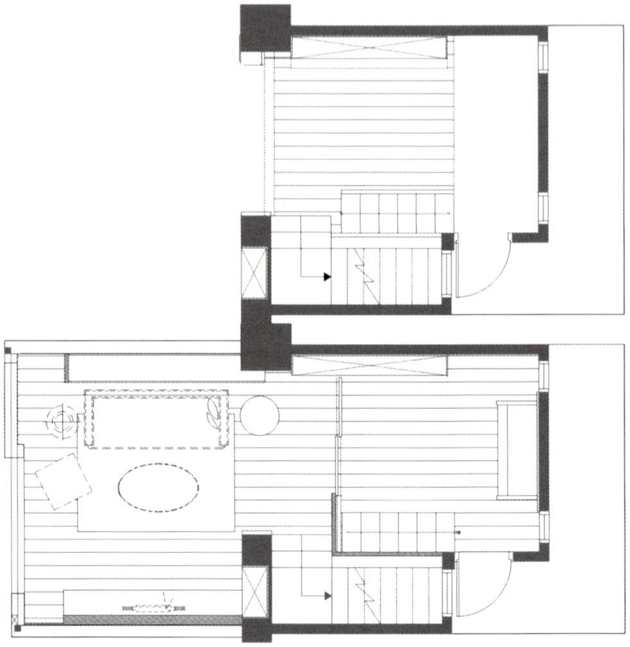

该建筑为独栋的透天别墅，整合了办公空间与居住空间，使其相互关联而又各自独立。一楼为办公空间，厨房与餐厅作为员工的会议区与聚餐场所，采用了纯白淡雅的乡村风格，细节精致细腻。楼梯将居住空间一分为二，隔出书房与卧室的场域。狭长走廊中的穿衣镜放大了空间维度，没有过多缀饰的天花板让视觉的重点放在优美的法式线板上，突显古典风味。四楼分别为男孩房与女孩房，一样的面积规划出不同风格的效果。顶楼休憩室为挑高的楼层，在电视墙与壁柜的比例上拿捏得恰到好处。整体色彩以灰、白为基调，辅以宝蓝色的落地窗帘与醒目的红色灯具，和谐而精致，将法式古典风格精美呈现。

室内设计

Lisa Lougee Interiors

景观设计

Loretta Gargan Landscape + Design

摄影师

Paul Dyer

Open Box House

开口箱房屋

Emerging from humble beginnings, Open Box was transformed from a run-of-the-mill 1950s vinyl-clad box into a contemporary family home that takes advantage of its site. The original house was detached on three sides and perched on a lot with panoramic views, but its small windows and dark rooms ignored the hillside location and potential for views. The unfinished lower floor made no effort to connect to the large undeveloped yard.

Interior walls were removed to create a large open plan that centers on a sleek new kitchen and floor-to-ceiling windows were installed to offer stunning downtown views. To the previously unfinished lower floor a new entry was added, along with a guest suite and a large family room where glass walls roll away to connect the inside to a newly landscaped deck and yard. The clean modern house is warmed by the generous use of dark stained woods and rich tiles and by its strong connection to lush landscape. Steel and aluminum accents and details contribute to its clean design.

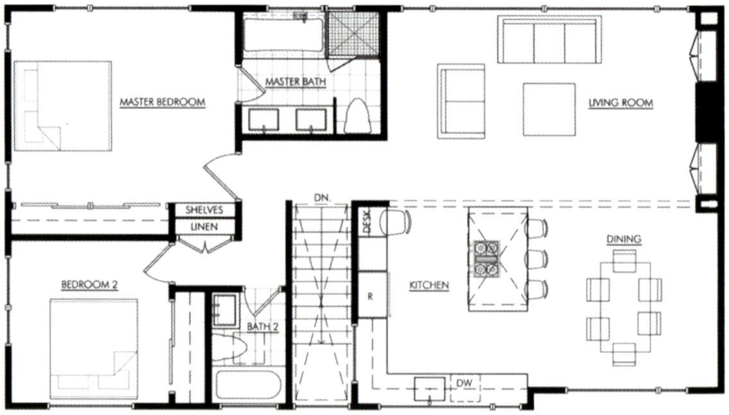

SECOND FLOOR

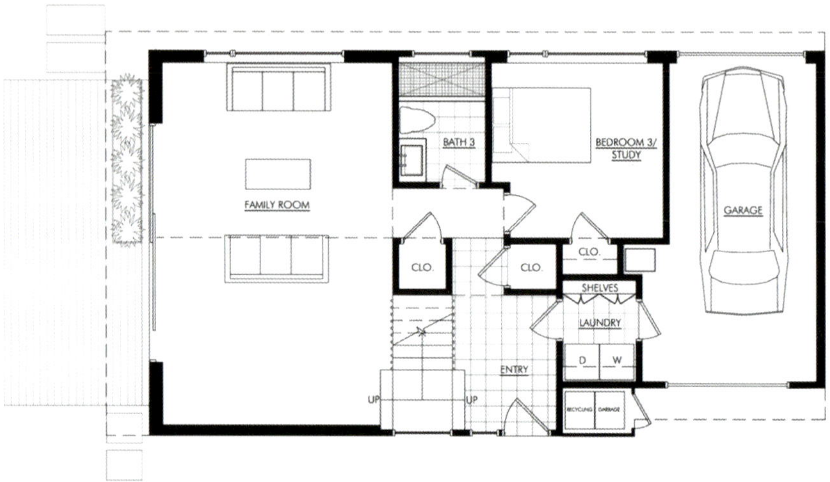

FIRST FLOOR

FABERGÉ

设计将这栋朴实的建筑从20世纪50年代平凡的乙烯基覆层箱子变成一个现代的家庭住宅，并利用了其位置优势。原来的房子分为三面，栖息在一块空地上，但其小小的窗户和黑暗的房间削弱了有利的山坡位置和取景潜力。未完成的较低层对大面积未开发的院子根本起不到连接作用。

设计师拆除了室内墙壁，营造出大型开放式空间，中心是光亮的新厨房，安装落地窗以供人们欣赏令人惊叹的城区风景。在之前未完成的较低层上增加了一个新入口，玻璃幕墙沿着客房和大型家庭房伸展，连接着室内与新观景平台和院子。深色的染色木料和大量瓷砖的大气运用，与郁郁葱葱的景观紧密相连，温暖了这个干净的现代房屋。钢与铝的质感和细节突出了其简洁的设计。

地点	面积	设计师	设计公司	主要材料
昆明	500平方米	何小龙	昆明中策装饰有限公司	大理石、壁纸、地毯、木地板

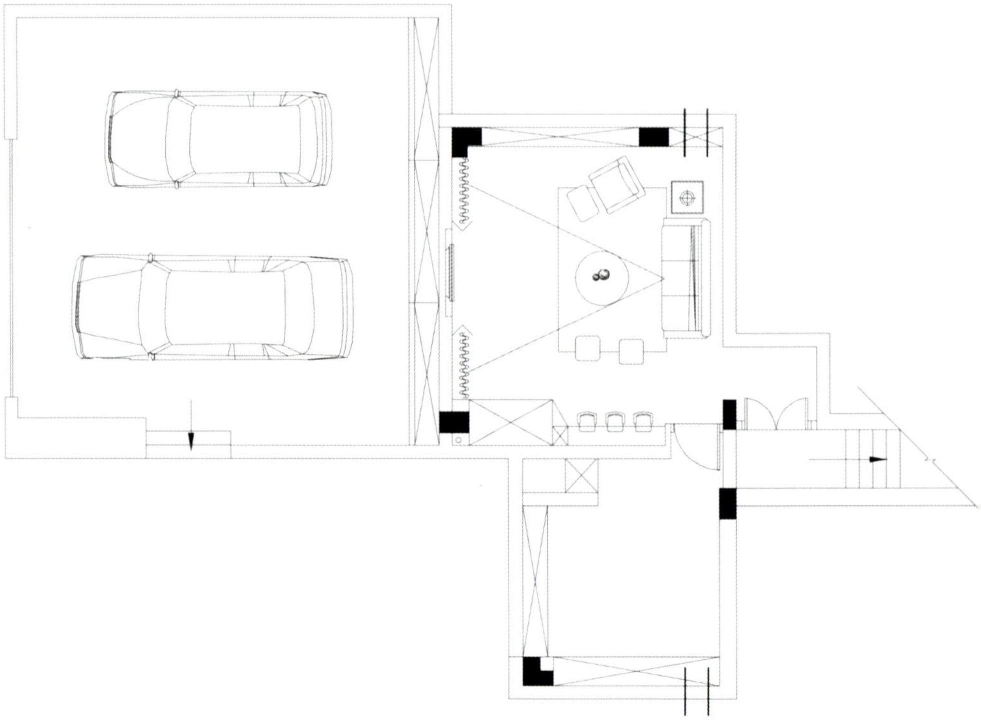

Dianchi Lake Golf Mansion

滇池高尔夫豪宅

This is a space that allows the designer to give full play, and the structure is very regular with clear-cut distinction between the rest area and the activity area. The whole design is mainly simple European style, coupled with some red that less common in simple European style. The introduction of red makes the whole space no longer serious and cautious, otherwise, tending to a little lively and agile. Many places are designed with French windows to improve the lighting, echoing with the theme of large space. Therefore, the designer's unique idea can be seen. The design carefully selected the patterns of the floor and steps, though the patterns seem to by fancy, for the coordination of the colors and overall environment, they do not look so awkward.

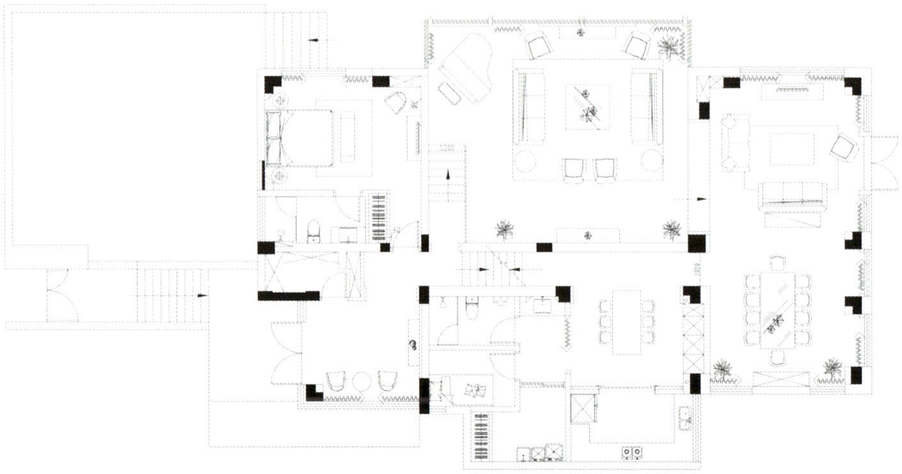

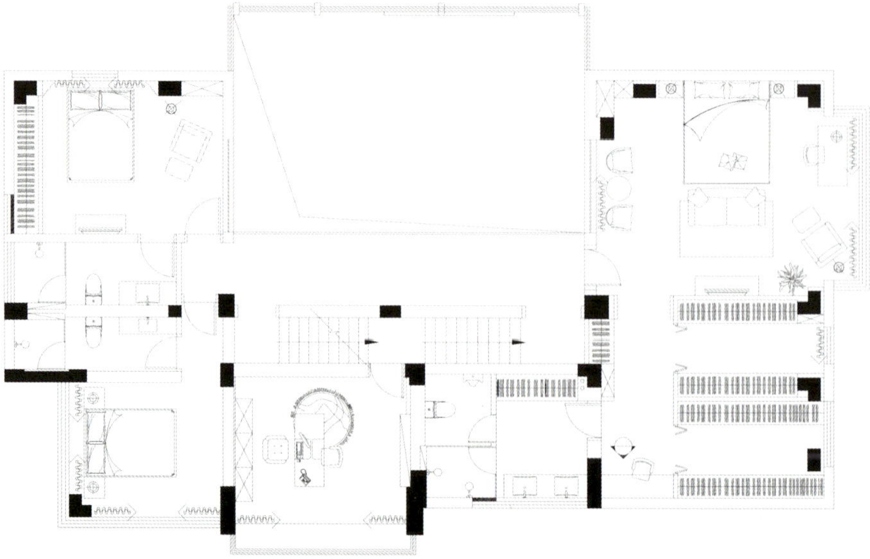

这是一个能让设计师充分发挥的空间，格局很规整，休息区与活动区被干净利落地区分开。整个设计风格以简欧为主，并加上一些在简欧中不太常见的红色来搭配。红色的加入，使整个空间不再严肃拘谨，倾向于些许的活泼和灵动。多处的落地窗设计，让采光性大为提高并和大空间的主题相呼应，设计师的独特想法于此可见一斑。地板和台阶的花纹选取也经过斟酌，看似花哨的图案，却因为颜色和大环境的协调性而不显得那么突兀。

地点	面积	设计师	结构设计	摄影师
El Campello (Alicante), Spain	204平方米	Moho Architects (Carlos Abadía Suanzes-Carpegna, Ignacio Bautista Ruiz, Pablo García Mora, Ernesto Sierra Díaz) + Ana Gilsanz Díaz	Abelardo Granados	David Frutos

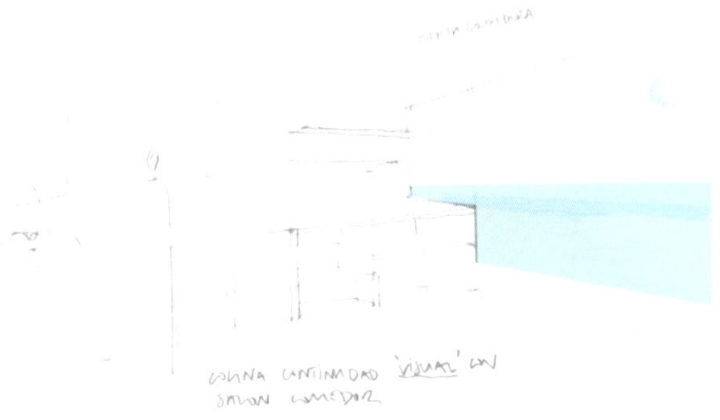

Apartment in Alicante
阿里肯特公寓

The reform involves the conversion of two apartments in a duplex apartment, occupying the second and third floor of a residential apartment building in front of the Mediterranean Sea in Alicante (Spain).

With the prevailing conditions given as the structure (at the time of placing the staircase to connect both levels), available wet rooms, research for the best direct view over the sea and need to create multiple storing spaces, the designers created a clear distribution scheme where the conventional scheme of housing is reversed, with the main access by the top floor.

At this level, the main spaces available during the day are placed around a "wooden box" which, running through the corridor of access, connects the staircase with a large dining and living room with great views to beach. It also houses the toilet and kitchen areas.

On the lower floor is where the “night” and most private spaces are placed: A bathroom and three bedrooms, one of them, the master bedroom, connected to a dressing room, a bathroom and a study.

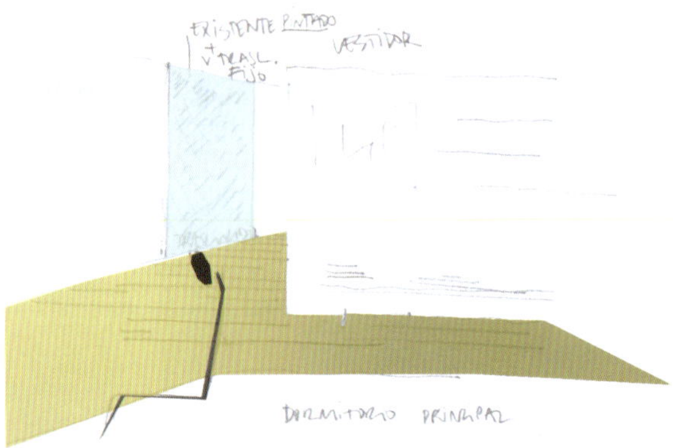

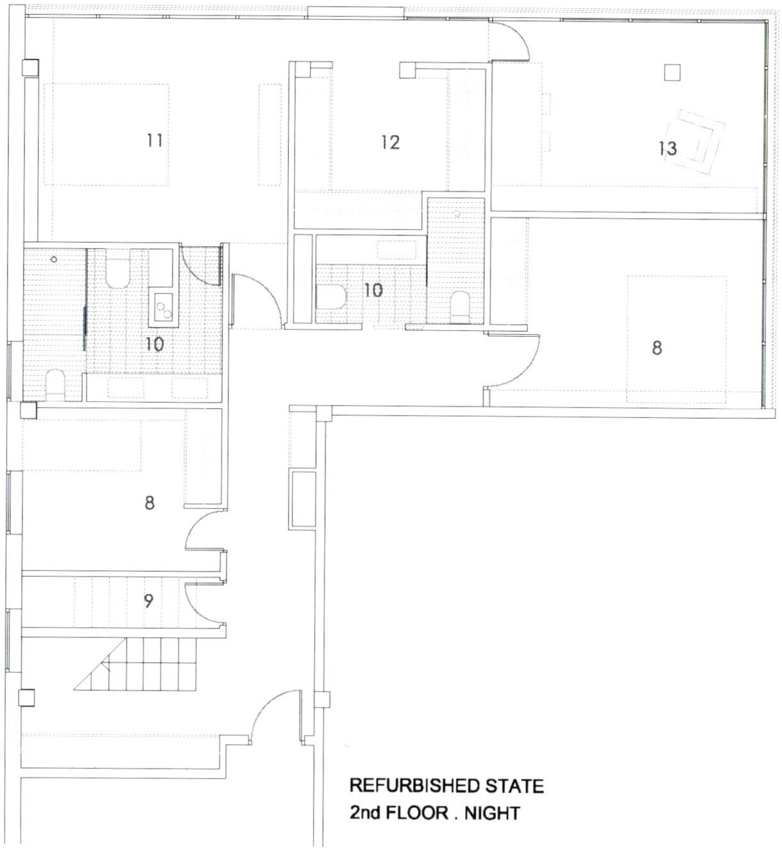

REFURBISHED STATE
2nd FLOOR . NIGHT

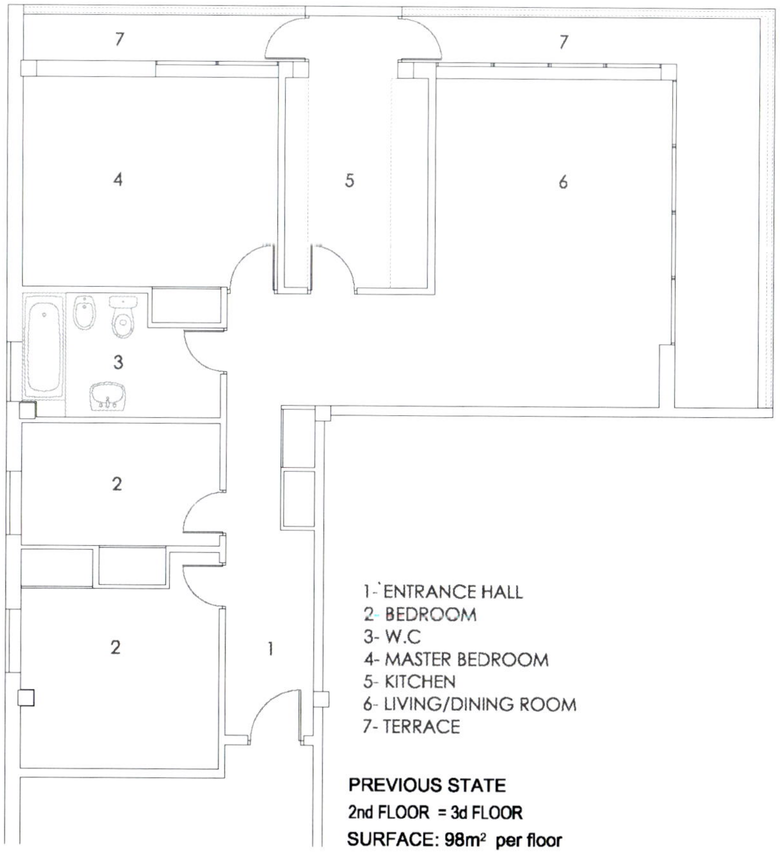

1- ENTRANCE HALL
2- BEDROOM
3- W.C
4- MASTER BEDROOM
5- KITCHEN
6- LIVING/DINING ROOM
7- TERRACE

PREVIOUS STATE
2nd FLOOR = 3d FLOOR
SURFACE: 98m^2 per floor

1- ENTRANCE HALL
2- TOILET
3- GALLERY
4- KITCHEN
5- DINING ROOM
6- LIVING ROOM
7- TERRACE
8- BEDROOM
9- CLEANING STORAGE
10- WC
11- MASTER BEDROOM
12- DRESSING ROOM
13- STUDY

REFURBISHED STATE
3 FLOOR . DAY

本项目位于西班牙阿里肯特地中海海滨的一座公寓建筑中，此次改造的是公寓第二层和第三层的两间。

鉴于建筑构造的普遍状况（当时是以楼梯来连接上下两层）和浴室的情况，结合对最佳海景的位置选取以及营造多储物空间的需求，设计师们打造了一种清晰明了的分配方案，保留了传统的住宅设计方案，并将重点放置于了顶层。

顶层是白天活动的区域，主要空间围绕一个“木箱”展开，穿过走廊，楼梯连接拥有无敌海景的宽敞餐厅和起居室。卫生间和厨房也被设置在这一层。

“夜晚”空间和大多数私密空间被安置在了下层：包括一间卫生间和三间卧室，其中有一间主卧室，连着更衣室、卫生间和书房。

地点	面积	设计师	设计公司	主要材料
东莞	389.8平方米	董晋强	广东震旦建筑设计有限公司 东莞市大木建筑设计有限公司	乳胶漆、白橡木、实木地板、新沙安娜米黄

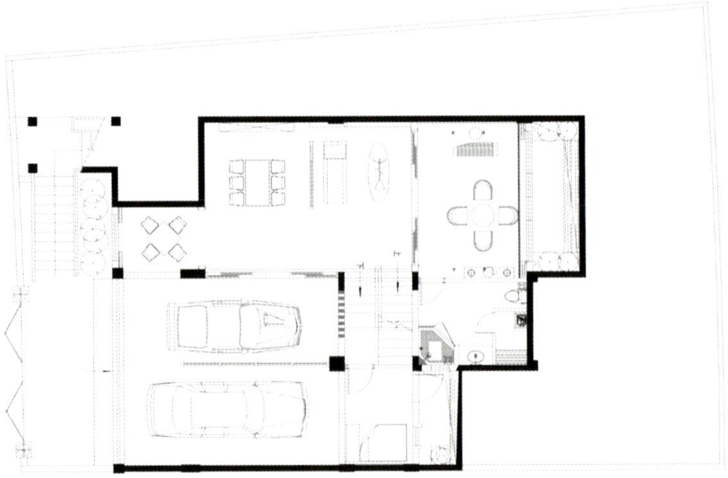

Wolong Mountain Garden Emerald Style D Sample Villa, Fenggang

凤岗卧龙山花园翡翠D型别墅样板房

The design of this project is mixed with different styles, neo-classical, Southern Californian, and new Chinese; luxury and generousness are the highlight of the design. In the living room, the elegant green is coupled with gold to make the space elegant and high-profile, plain and cute. The space is mainly warm tone, and the delicate handling of details and use of different colors make the space integrated and rich. The open-style pattern and introduction of large area of light colors greatly stretch the vision in the space, generous without losing the integrity of the layout. The Chinese-style dining chairs are coupled with European-style droplight, fabric sofa is matched with wooden furniture, and the wood color is combined with white color, so the space fully shows the beauty of harmony in the mixing style. The complicated but orderly curly grass wallpaper endows the space with much rhythm, enriching the visual levels in the space.

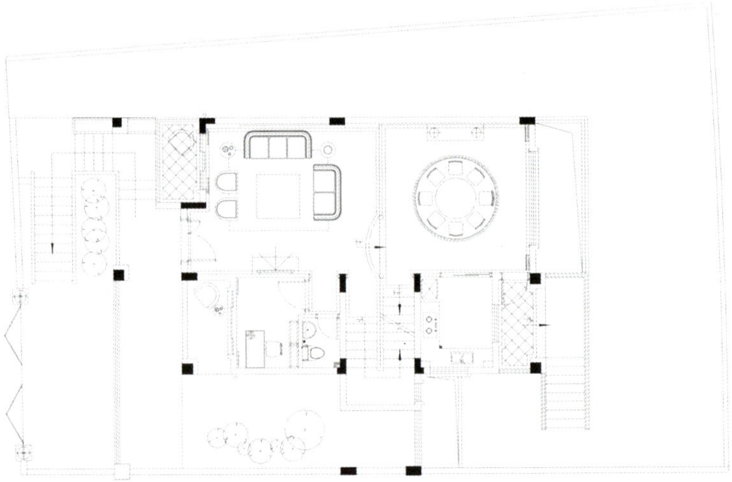

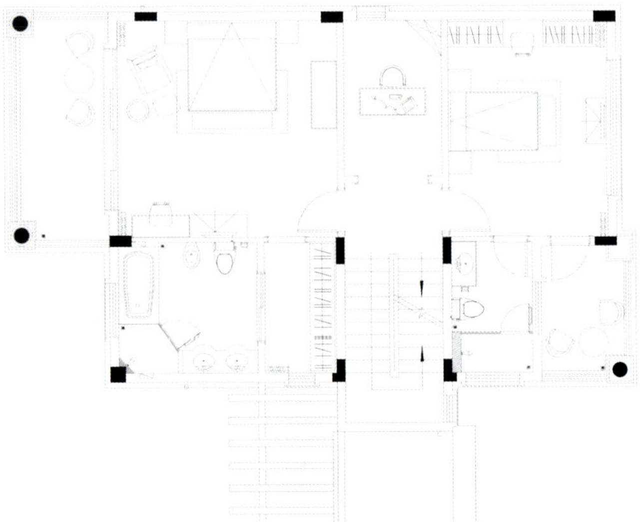

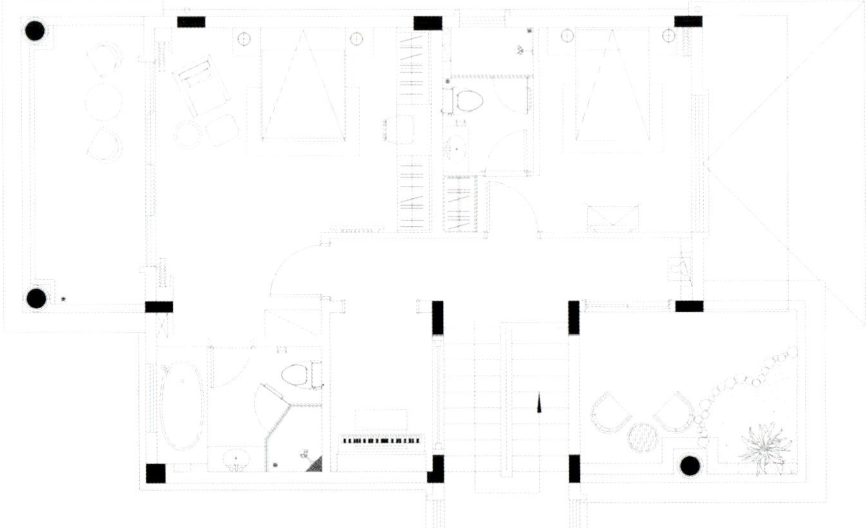

USTYLE
TM420

在本案设计中能看到新古典、南加州、新中式等不同风格的混搭之风，豪华与大气是本案设计中强调的重点。客厅中淡雅的绿色与金色的搭配，让空间变得优雅中不乏高调，平淡中不乏俏皮。空间以温馨的暖色调为主，通过细节的精巧处理和色差的运用，让空间变得完整又丰富。开放式的格局和大面积浅色的运用，极大地舒展了空间的视线，大气又不失布局完整。中式的餐椅与欧式吊灯的搭配，布艺沙发和木质家具的组合，原木色与白色相间，无不让空间在混搭之中尽显和谐之美。繁复而又有序的卷草壁纸为空间增添了不少韵律与节奏，让空间的视觉层次丰富了不少。

地点	面积	设计公司	摄影师
日本爱知县津岛市	87.73平方米	森田一弥建筑设计事务所	渡边慎一

Pentagonal-house

五角住宅

Think about roofs, people can say it is another landscape, which is created from where the architecture would build. As well as the natural landscape has, they are full of attractiveness for the space to live for people. And at the same time, roofs had formed by integration of different kind of technologies, so people can say roofs as "meta-technology" of architecture. If people start designing architecture from thinking about roofs as meta-technology, they have chance to deliver another technology for making roofs to be better landscape to live.

The site is within a calm village near Nagoya, Japan. It is a house for young couple, and just next to this architecture, there is an old Japanese style house their parent lives. To respect for and harmonize with neighboring environment, the designers delivered traditional hipped roof as many neighboring houses. This characteristic pentagonal geometry of plan was delivered to make the maximum space in this site and adequate open space around it. Then the designers start to think about how to live under the second landscape, pentagonal hipped roof.

Five main structural walls were set in radially, which makes possible to take in the outside spaces as extension of interior spaces. In the center of the architecture, radial walls were cut off in dome shape, to make space for a dining table. Here, under the peak of the landscape, the designers have tall ceiling height and whole families can enjoy their dishes all together and wide range views to gardens. In 5 surrounding spaces with low ceiling height, under the skirts of the landscape, people can stay calm and relax with appropriate distance from others, just like a life in the traditional house.

Besides delivering this characteristic geometry (it is a kind of technology) to the plan of this architecture, the designers tried to use the most usual wooden structure system what most Japanese houses are constructed nowadays. Walls were finished with round corner in Japanese traditional white plaster.

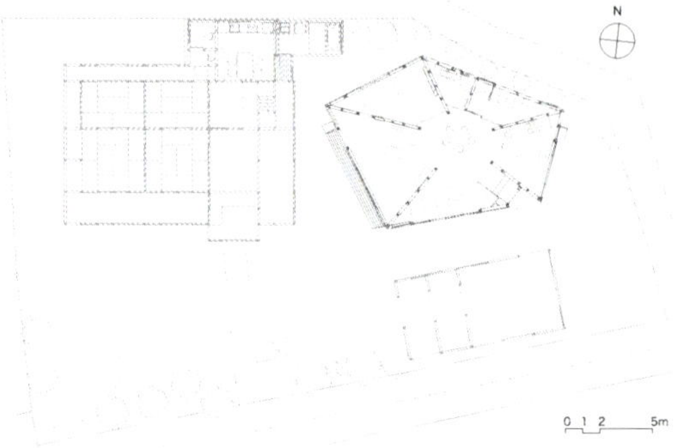

一想到屋顶，人们可能会说是由建筑创建的另一种景观。除了拥有自然景观，它们还吸引着人们进入空间来生活。同时，屋顶由不同类型的技术整合而成，所以可以称屋顶为建筑的“中继技术”。如果将中继技术应用到屋顶的设计中，就有机会为创造更好景观生活的屋顶而传播另一种技术。

本案位于日本名古屋附近的平静村庄。这是一对年轻夫妇的住房，附近就是其父母居住的古老日式风格的房子。为尊重并与周围环境相协调，本案沿用了附近很多房子使用的传统四坡屋顶。设计这个典型的几何形五角设计可以使其内部空间最大化，并充分利用其周围空间。然后，设计师们开始考虑如何在第二层风景——五角形四坡屋顶下生活。

五个主要结构墙呈放射状设计，使外部空间可以延伸至屋内。在建筑的中心，放射状墙体被切割成圆形，腾出空间摆放餐桌。在这里，在风景如画的山峰之上，设计了高大的天花板，全家可以共同享受美食及花园美景。周围设计了五个低天花的空间，在美景下，人们可以冷静、放松，与其他人保持适当距离，就像生活在传统房子里一样。

除了在本案中使用这种特性几何（它是一种技术），人们还试图使用现在日式房屋建造中最常见的木质结构系统。墙壁上利用传统的日式白石膏做成圆角。

地点	面积	设计师	设计公司	主要材料
福州	280平方米	陈住光	福州好日子装饰工程有限公司	道格拉斯仿古砖、世纪米黄大理石、沙比利面板、茶镜、砂岩、天伦墙纸、软包

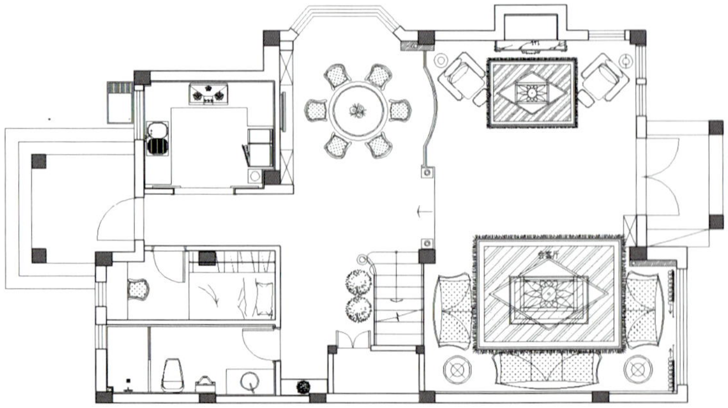

Southern County of Lake Gui, Fuzhou

福州桂湖南郡

This project has a very magnificent spatial hall, the style of which is elegant, graceful and refreshing. The designer elevates the area between the living room and the dining room, and designs exquisite fences, so that these two parts are interlinked but not connected, meanwhile making the vision extended. Two luxury chairs and a soft carpet are placed beside the fireplace, which become the central focus, giving off a sense of coziness and comfort. The highly textured stone flooring, the luxuriant furniture arrangement and precious arts collections: all can provide a glimpse of the owner's noble life. In such a peaceful project, one can take a rest, and have a relaxation, enriching the life in a rich fantasy and fairness.

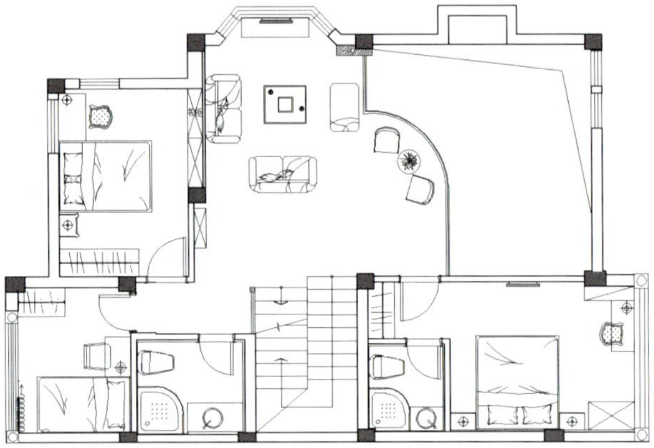

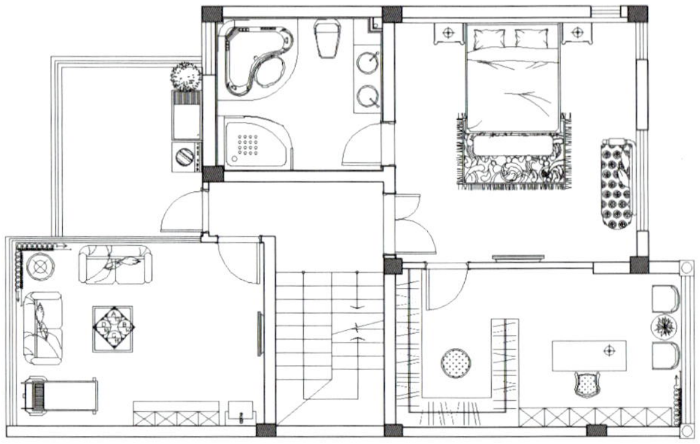

本案拥有非凡气派的挑空大厅，风格典雅优美、飘逸灵秀。设计将客厅与餐厅区域之间做抬高处理，并安装了精美的栏杆，使这两部分空间相连而不相通，并使视觉得以延伸。两把华贵的椅子与一张柔和的地毯布置在客厅壁炉旁，形成了视觉的中心点，给人以温馨之感。从充满质感的石材铺地、华丽的家具布置以及名贵的艺术收藏品中，可一窥业主贵族式的生活。业主可于此悠闲品茗，放松心情，在浓厚的梦幻与美丽之中丰富人生。

地点	面积	设计师	设计公司	主要材料
哈尔滨	508平方米	万文拓、彭秋波	深圳市品源装饰工程有限公司	地毯、壁纸、大理石

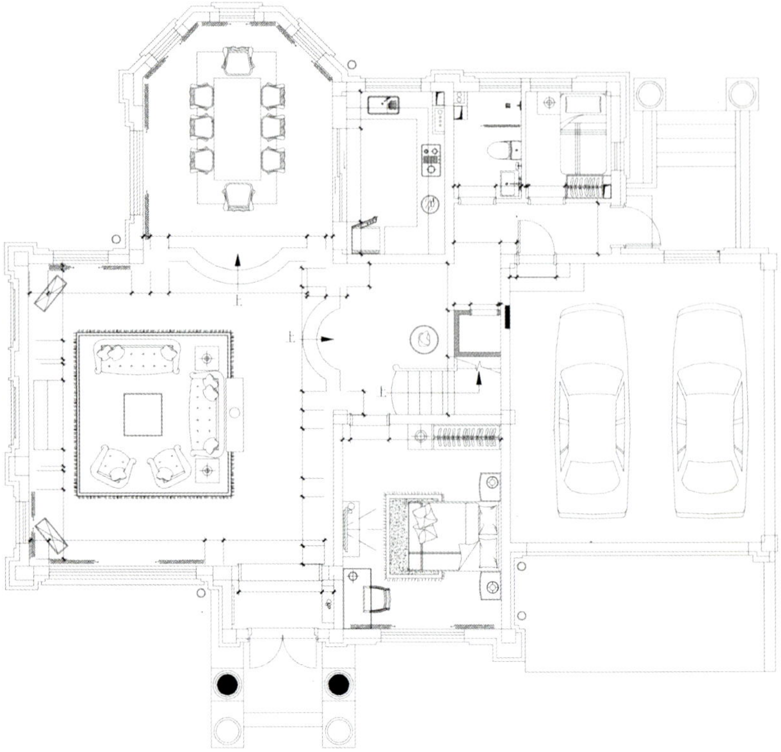

Russian Style of "Sunny Mansion"

"阳光府邸"俄罗斯风格

The appearance of the building is classical European style, and the interior design continues the style to keep consistent. The Russian style is introduced, and the designers make extensive use of European-style internal corner lines, Rome pillar, goldleaf engraved ornamental lines, wall panels, fireplace and other elements. The ceiling of the living room is embellished with goldleaf decoration, and the floor is paved with light yellow and brown marble, unobtrusively showing out the sense of luxury. The bedrooms and study are designed with beige-color wallpaper and warm-color lights, blowing warm sense to the visitors. The overall is white and gold, generous, stable and luxurious. The golden crystal droplights and wall lamps make the whole space luxurious, meanwhile, bringing some mellow flavor to the space.

本栋建筑外观为欧式古典风格，室内装饰亦延续了外观的风格。设计上采用俄罗斯风格，大量采用了欧式的顶角线、罗马柱、金箔镂花装饰线、护墙板、壁炉等元素。在客厅吊顶中，金箔装饰点缀其中，而客厅地面浅黄色和咖啡色大理石相拼，不张扬地将奢华感体现出来。卧室、书房等处选用米色系的壁纸和暖色系的灯光，温馨感扑面而来。设计整体为白色和金色，大方、稳重、奢华。空间中的一盏盏金色的水晶吊灯和壁灯，让整个空间变得豪华起来，同时多了些许柔美的韵味。

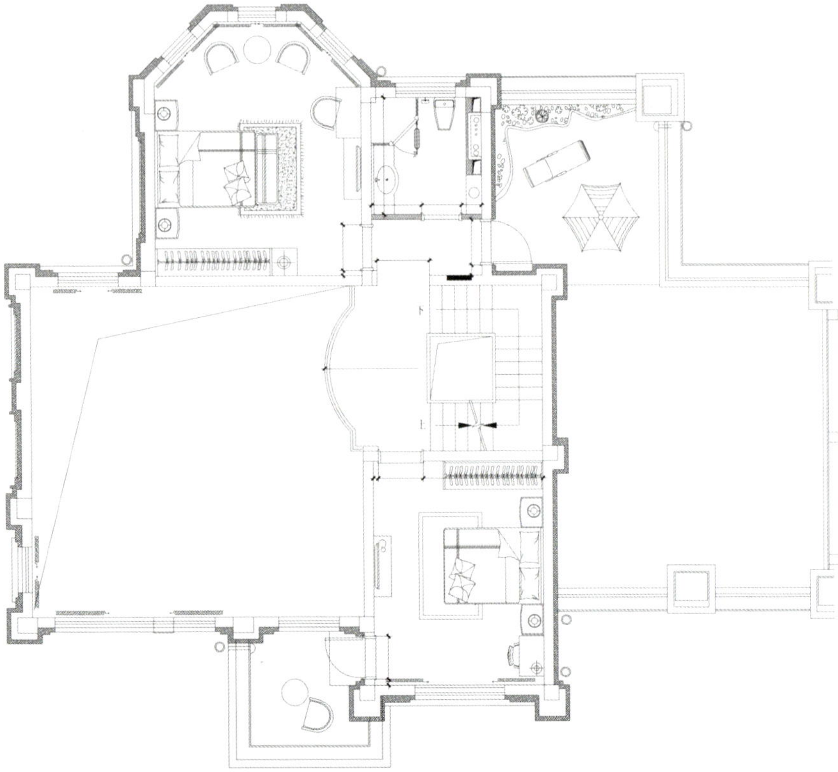

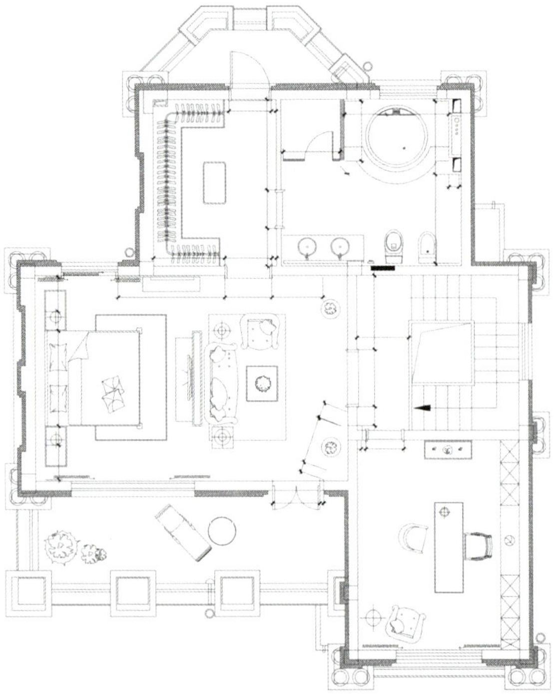

地点	面积	设计师	设计公司	主要材料
北京	270平方米	赵巍、刘鑫	北京山禾金缘艺术设计有限公司	瓷砖、石材、地板、壁纸

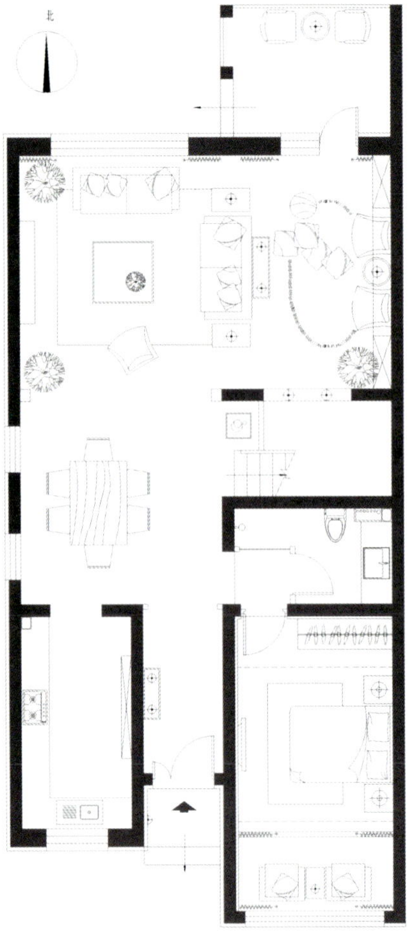

Tiziano Villa D3

提香草堂D3别墅

This project takes the European style as the main design style, which gives prominence to the changes of moving lines, and is fabulous in colors. In terms of forms, this project is based on the romanticism. In regard of the decoration materials, it employs marble, colorful fabric, fancy carpets, and exquisite wallpaper, so that the whole style is luxurious and grand. The appliance of curves turns the effect to be more elegant and gracious.

Set off by the orange colored curtains, the master bedroom appeals much warmer. This rest area, though not very large, has integrated functions. How pleasant and comfortable it is to lie on the round bed enjoying the natural and dusky sunlight. The sedate dining table and exquisite lamp create a peaceful dining atmosphere imperceptibly. The design of this space is not merely luxurious and grand, but more pleasant and romantic. Through the perfect curves and the excelsior treatment of details, it has brought endless enjoyment.

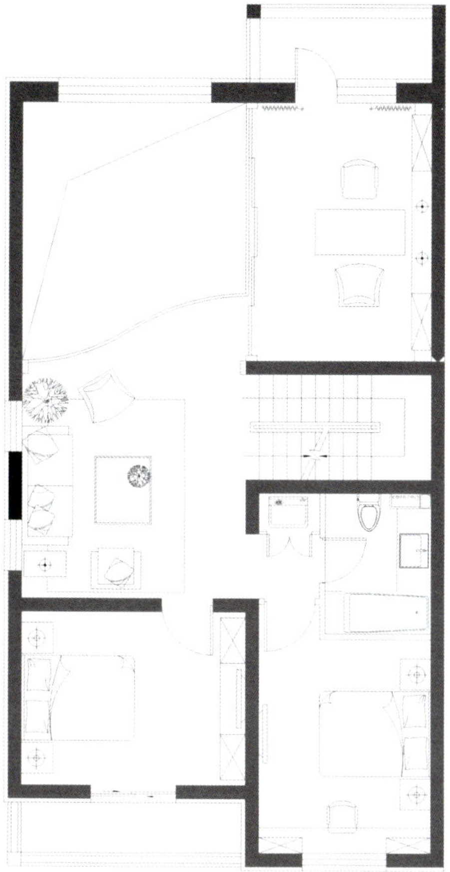

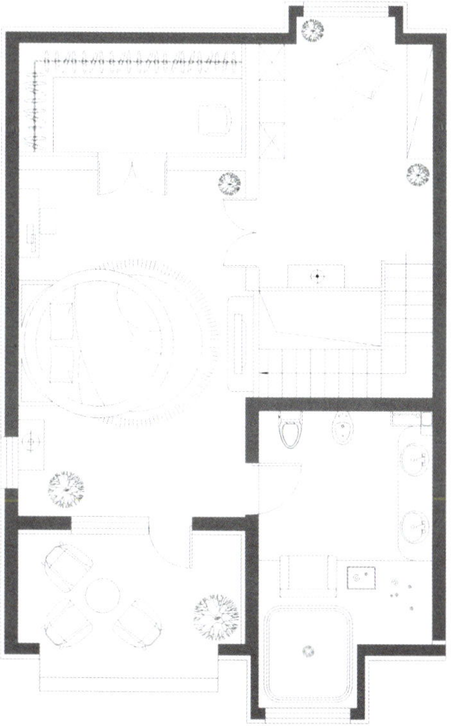

本设计主要以欧式风格为主，强调线性流动的变化，色彩华丽。它在形式上以浪漫主义为基础，装修材料运用了大理石、多彩的织物、精美的地毯、精致的壁纸等，整体风格豪华富丽。曲线的运用，使效果更加典雅、亲切。

主卧室在橘黄色帘幕的衬托下，显得更加温暖，不大的休息区却有着完整的功能。躺在圆形床上享受自然朦胧的阳光是多么的惬意。稳重的餐桌和精美的灯饰在无形中营造了一种平静的就餐气氛。空间的设计不只是豪华大气，更多的是惬意和浪漫，通过完美的曲线，精益求精的细节处理，带给人无尽的享受。

地点	面积	设计师	设计公司	摄影师
Talstrasse 175 A - 3491 Strass	650平方米	Denis Kosutic	architektur denis kosutic	Lea Titz

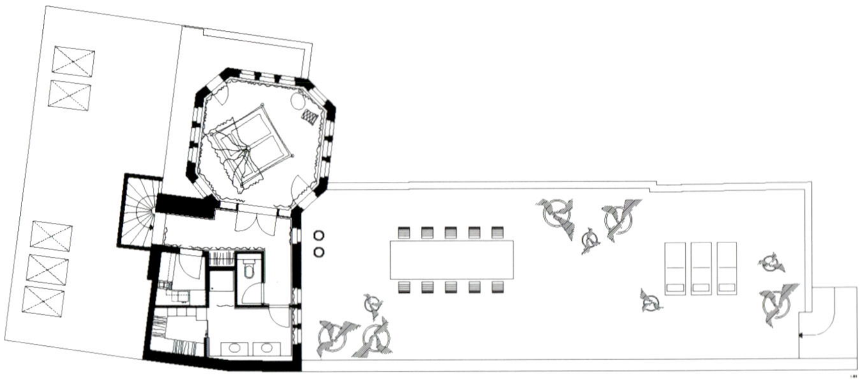

CASTLE S

S城堡

The functional room framework of the three-storey house has been partly adopted, partly modified and partly newly created. The countless small or large rooms thereby have been meticulously dealt with separately. The various moods and sources of inspiration translated into the design of the individual rooms are always independent from each other and yet familiar.

A wild, bold and haphazard mixture of old and new serves as dominant element of the design. In this process, the "old" would constantly be newly designed—as a possible reconstruction or, as in most cases, as an artistic interpretation. An intentional renouncement of clear definitions, along with a casual play with the material both result in a "mix and match" of shapes, surfaces, colors and ornaments. An enormous quantity of wall claddings, ceiling panels, stucco, wallpapers, fabrics and antique carpets in compositions that appear randomly but are precisely attuned, making a warm and familiar, but at the same time surprising and unexpected effect. An abundance of antique, restored and newly designed furniture from all epochs have been brought into a dialogue with contemporary design and art objects, at times in dignified, at times in ironic compositions. The boundaries between old and new thereby totally disappear, creating a new, inherently consistent composition, which is open for all possible influences and changes in the future.

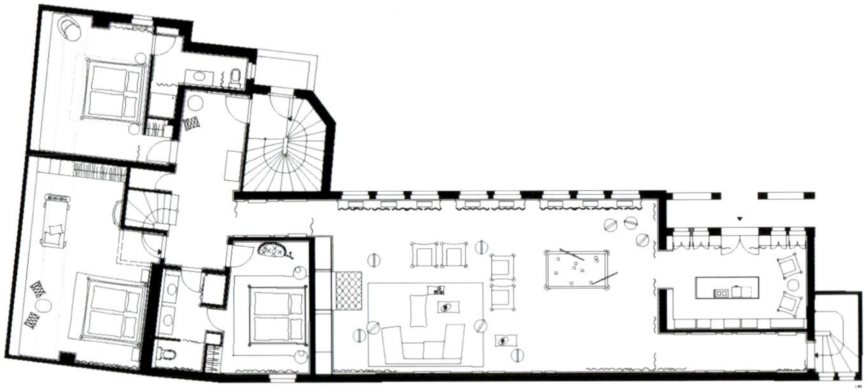

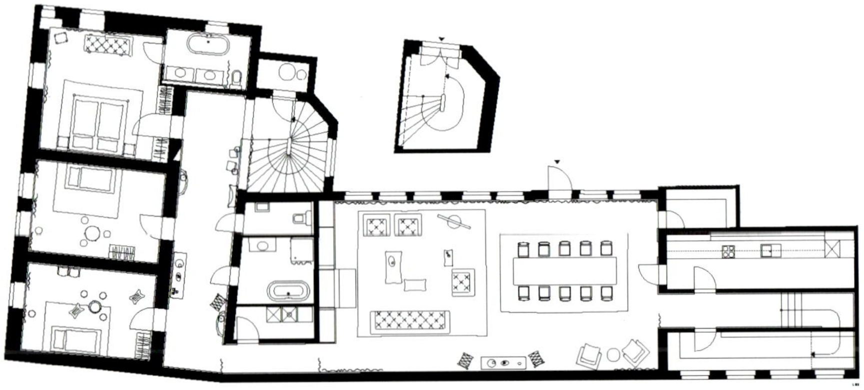

整个三层建筑功能房间结构的一部分沿用了原始结构，一部分为改造而成，而一部分则为新建造的。因而营造出的无数或大或小的房间都经过了特定的精心设计。设计师将各种情感和灵感来源转化成每个空间的设计，相互独立而又有些相似之处。

设计将新旧事物的狂野、大胆、随意混合作为该设计的主要元素。在设计过程中，“旧”事物被重新设计——有可能是重建，或在大多数情况下，采用一种艺术的诠释。设计师故意放弃了明确的定义，不经意间使用材料形成形状、表面、色彩及装饰品的“混搭”。无数墙壁面板、吊顶板、粉饰灰泥、壁纸、面料和古式地毯组合在一起，虽是随意呈现，却又如此地协调，营造出一种温暖而熟悉，同时令人惊讶而意想不到的效果。设计师将产于各个时代的古董、修复或新设计的家具与当代设计和艺术品互相搭配，营造出时而严肃、时而奇特的艺术效果。

从而，新旧的界线完全消失，整个设计营造出全新的、内在一致的组合，适用于未来可能产生的所有影响和变化。

地点	面积	设计师	设计公司	主要材料
无锡	486平方米	彭政	香港彭政设计师（上海）有限公司	米黄大理石、啡网纹大理石、银箔、真皮软包、复古砖、茶镜、灰镜、墙纸、马赛克、白漆等

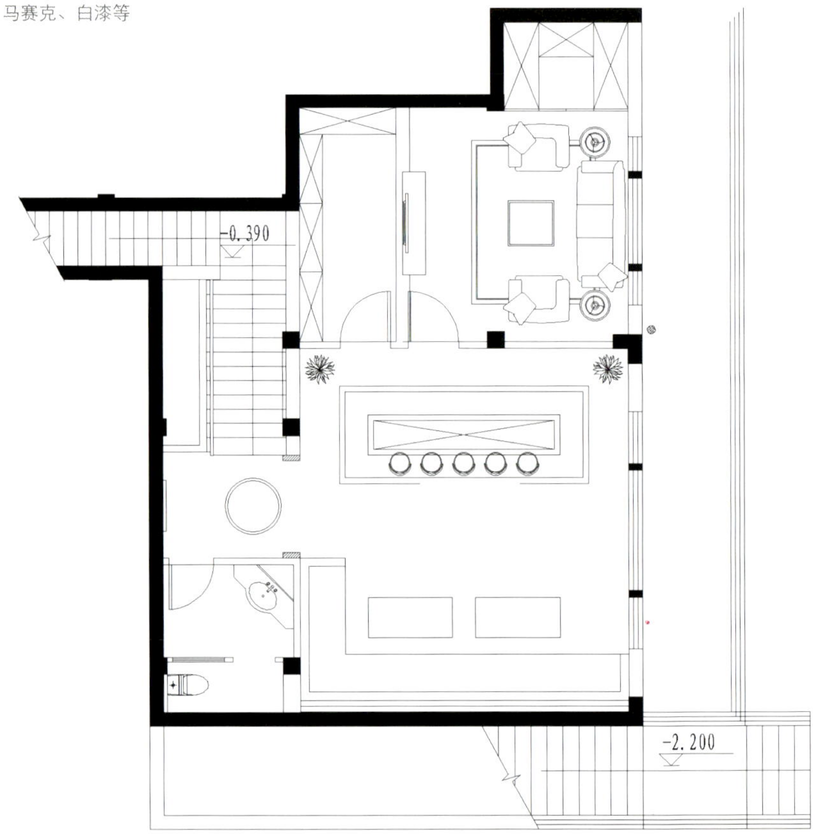

Tanxi Bay Sample Villa, Wuxi

无锡檀溪湾别墅样板房

The project is American neo-classical style. As reception area, the living room is bright and simple, and its decoration is more lively and bright than other spaces, designed with a large area of carpet accompanied by American-style sofa of elegant and comfortable shape. The floor curtains on both sides in the living room show the generous and luxurious atmosphere, and the crystal chandelier is absolutely a great highlight in the living room. In the dining room, the tableware is placed orderly and tidily, the transparent goblets, gorgeous table cloth, candlestick as a symbol of classic style, reminding people of the nostalgia feeling. The designer places the used antique in the most eye-catching position in the room, to show the atmosphere of culture and history. The design of bedroom is warm and fragrant. As the owner's private space, bedroom mainly focuses on the function and practical comfort. The archaized bedside lamps and photo frame are placed on the bedside cabinet, full of classical flavor. In the color matching, warm color is introduced, echoing with the light and elegant wall.

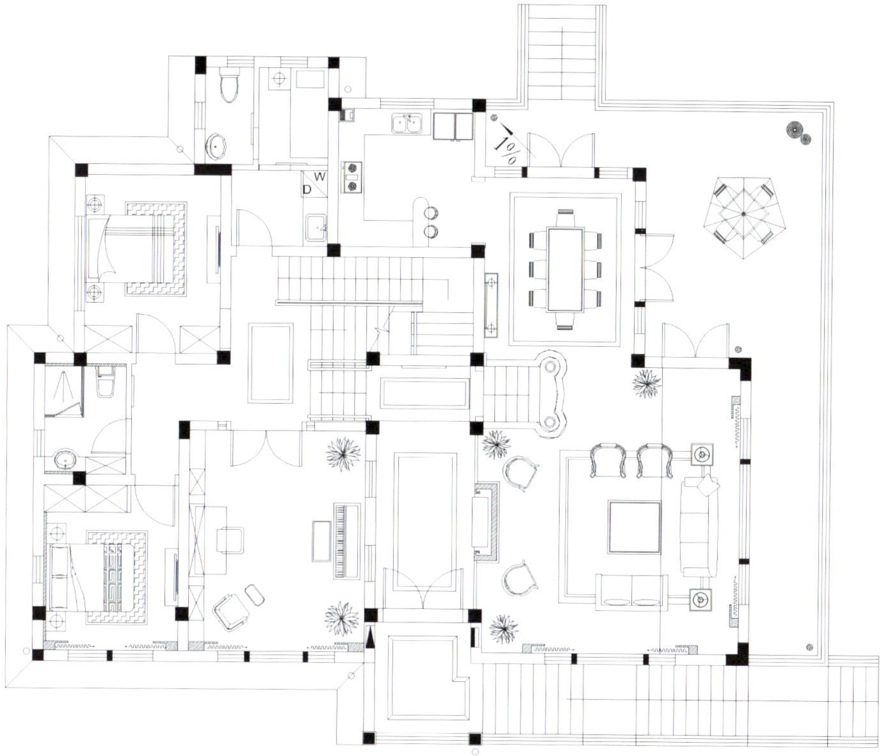

D
W

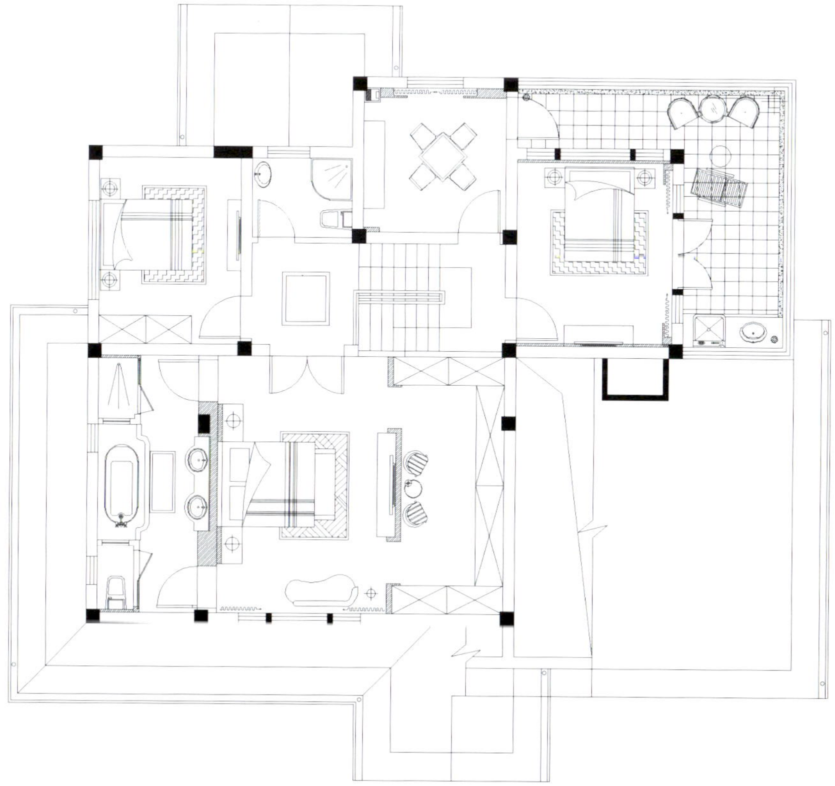

本案为美式新古典主义风格。客厅作为待客区域，简洁明快，装修较其他空间显得更明快光鲜，使用了大面积的地毯，并配以造型优雅舒适的美式沙发。客厅两面整体的落地窗帘，尽显大气华贵。水晶吊灯是客厅的一大亮点。餐厅中整齐有序的餐具摆设，透明的高脚杯，华丽的餐布，象征古典主义风格的烛台，让人产生身临其境的怀旧情绪。设计师将用过的古董摆设在居室最醒目的位置，从而显示文化和历史的风韵。卧室布置较为温馨，作为主人的私密空间，主要以功能性和实用舒适性为考虑的重点。设计师将仿古床头灯和相框摆设在床头柜，很有古典主义的风味。在色彩搭配上，设计师采用了暖色调与淡雅的墙面相呼应。

索引

时代花园6层别墅（A3栋）
设计师：刘伟婷
设计公司：刘伟婷设计师有限公司

文华酒店之东方魅力A1户型
设计师：刘伟婷
设计公司：刘伟婷设计师有限公司

香蜜湖48#墨西哥风情
设计师：史林艳
设计公司：香港神采设计建筑装饰总公司
宁波神采装饰设计工程有限公司

惠州保利山水城别墅
设计师：薛峰、朱晓鸥、王君明、娄欣
设计公司：寅界（国际）建筑室内设计

重庆小泉森林海别墅
设计师：冉旭、屈慧颖
设计公司：重庆旋木室内设计有限公司

台中市美术馆园道周公馆
设计师：吴玟灿
设计公司：夏川设计/夏川榭环境艺术设计有限公司

紫汀花园
设计师：朱森钥
设计公司：宁波市鄞州风尚装饰设计工程有限公司

“黑白映像”现代风
设计师：万文拓、彭秋波
设计公司：深圳市品源装饰工程有限公司

迷情法兰西
设计师：范义峰、黄婷婷
设计公司：上海天鼓装饰设计有限公司

杭州西湖新古典主义
设计师：梁苏杭
设计公司：东易日盛装饰集团杭州分公司

佘山的梦想
设计师：萧爱彬
设计公司：萧氏设计

紫檀宫
设计师：陈志斌
参与设计：李智勇、谭丽、彭辉
设计公司：鸿扬集团陈志斌设计事务所

海尔天津别墅A户型
设计师：范义峰、杨俊、黄婷婷
设计公司：上海天鼓装饰设计有限公司

“骑士时尚”瑞士风
设计师：万文拓、彭秋波
设计公司：深圳市品源装饰工程有限公司

山顶住宅
设计师：Safdie Rabines Architects

灯塔
设计公司：Linebox Studio

恩卡那森
设计师：Nacho Polo

珀丽湾别墅
设计师：何永明
设计公司：广州道胜设计有限公司

布埃纳·维斯塔
设计公司：Feldman Architecture with Joe Oroza Design
建筑商：Lara Construction

中央悦城别墅
设计师：王冠
设计公司：深圳市矩阵室内装饰设计有限公司

斯通宁顿
设计师：MARYANN THOMPSON, FAIA
设计公司：MARYANN THOMPSON ARCHITECTS

完美整合
设计师：张馨
设计公司：京轩空间设计/张馨室内事务所

开口箱房屋
室内设计：Lisa Lougee Interiors
景观设计：Loretta Gargan Landscape + Design

滇池高尔夫豪宅
设计师：何小龙
设计公司：昆明中策装饰有限公司

阿里肯特公寓
设计师：Moho Architects (Carlos Abadía Suanzes-Carpegna, Ignacio Bautista Ruiz, Pablo García Mora, Ernesto Sierra Díaz) + Ana Gilsanz Díaz
结构设计：Abelardo Granados

凤岗卧龙山花园翡翠D型别墅样板房
设计师：董晋强
设计公司：广东震旦建筑设计有限公司
东莞市大木建筑设计有限公司

五角住宅
设计公司：森田一弥建筑设计事务所

福州桂湖南郡
设计师：陈住光
设计公司：福州好日子装饰工程有限公司

“阳光府邸”俄罗斯风格
设计师：万文拓、彭秋波
设计公司：深圳市品源装饰工程有限公司

提香草堂D3别墅
设计师：赵巍、刘鑫
设计公司：北京山禾金缘艺术设计有限公司

S城堡
设计师：Denis Kosutic
设计公司：architektur denis kosutic

无锡檀溪湾别墅样板房
设计师：彭政
设计公司：香港彭政设计师（上海）有限公司